Nineteenth-Century
SPECTROSCOPY

WILLIAM McGUCKEN

Nineteenth-Century
SPECTROSCOPY

Development of the Understanding of Spectra

1802-1897

The Johns Hopkins Press
Baltimore and London

The Johns Hopkins Press, Baltimore, Maryland 21218
The Johns Hopkins Press Ltd., London

Library of Congress Catalog Card Number 74-94886
Standard Book Number 8018-1059-0

To Walter F. Cannon

Contents

||||||||||||||||||||||||||||||||

Illustrations

Acknowledgments

This volume is, with a few minor emendations, my doctoral dissertation, written at the University of Pennsylvania, which I attended as a Thouron Scholar. For that privilege I am deeply thankful to Mr. and Mrs. John R. Thouron, C.B.E.

My interest in the history of spectroscopy was first aroused during a graduate seminar on physics from Thomas Young to Bohr. This was given by Dr. Walter F. Cannon of the Smithsonian Institution, to whom I am most grateful for his supervision of this book from its conception as a dissertation to its publication. All shortcomings, needless to say, are mine alone.

I am also indebted to Dr. Philip George and the Graduate School of Arts and Sciences, both of the University of Pennsylvania, for the award of a Dissertation Year Fellowship. This enabled me to complete much of the research and writing of the book.

Introduction

My story tells of the efforts made, and the success achieved, in understanding spectra during most of the nineteenth century—that is, from William Hyde Wollaston's observation of dark solar lines in 1802 to J. J. Thomson's discovery of the electron in 1897.

I begin by examining the origins of spectrum analysis as established in 1860 by Robert Bunsen and Gustav Kirchhoff. The central theme here is the early development of the principle of spectrum analysis, which states that each 'substance' has a characteristic spectrum. The further development of this principle in the 1860s and 1870s is discussed in sections 2, 3, and 4 of chapter II; and as this involves a significant change in physical notions of atoms and molecules, and consequently in spectral theory, an account of the original (molecular) theory of spectra is given in section 1. Indeed, the understanding of spectra at an atomic/molecular level forms the major theme of the remainder of the story. The final section of chapter II deals with Norman Lockyer's hypothesis of the dissociation of the elements, which was prompted by spectral considerations. This is a subject closely related to those of the previous sections of the chapter.

Spectroscopy, meanwhile, had led in a second way to an interest in atoms and molecules. It was thought that if the law or laws governing the distribution of lines in spectra were found, then something might therefore be known of the physical nature of atoms and molecules. The search for such a law or laws forms the subject of chapter III. Once such laws were found, and even before, the need for a mathematical theory of spectra was felt. Several attempts were made to give such a theory, but before considering these in chapter V, account has first to be taken of other developments in physical science which, either independently or in conjunction with spectral considerations, were also changing the physicist's conceptions of atoms and molecules. Accordingly, chapter IV considers the evolution of qualitative theories of spectra during the last quarter of the century. An interesting difference is found between the British and the Germans in their choice of an atomic model, and a link is established between the British model and Zeeman's discovery of the magnetic splitting of spectral lines.

In the epilogue, the discovery of the electron is seen as bringing to a close one era, while simultaneously ushering in another, in the continuing attempt to understand spectra.

I have not attempted to write a history of spectroscopy during the nineteenth century. No accounts will be found of such subjects as spectroscopy's role in discovering further elements, the rise of astrophysics, infra-red spectroscopy, radiation studies, spectroscopic apparatus, and many others. But where any subject becomes important for a full understanding of my story, I have made appropriate mention of it. Thus, for example, there is no discussion of the development of diffraction gratings, but when Henry Rowland's concave grating leads in the 1880s to a considerably higher degree of accuracy in wavelength determinations its importance is mentioned.

Throughout the work I have endeavored to get to the root or roots of each development, to know exactly how it occurred, and to be aware of its various connections.

Nineteenth-Century
SPECTROSCOPY

I

Origins of Spectrum Analysis

WHILE THE OCCURRENCE of dark, or Fraunhofer, lines in the solar spectrum was generally known in Europe from about 1820, their origin was not successfully explained until some forty years later. The successful explanation was given in 1859 by the Professor of Physics at Heidelberg, Robert Gustav Kirchhoff, who, about the same time and in collaboration with his counterpart in chemistry, Robert Bunsen, placed in order the many and complex spectra of incandescent salts. These splendid achievements formed the basis of spectrum analysis, whose origins in the progressive scientific climate of the nineteenth century are here considered.

The first section of this chapter tells how the Fraunhofer lines were discovered, and mentions some related observations of flame spectra. The next three sections treat main divisions of the story from the time of Fraunhofer down to the years immediately before Bunsen and Kirchhoff's collaboration—section 2 dealing with colored flame spectra in relation to chemical analysis, section 3 with the so-called electric spectra of metals, and section 4 with attempted explanations of the Fraunhofer lines. Section 5 considers the work of Bunsen and Kirchhoff and the immediate events leading up to it. The final section deals with Kirchhoff's explanation of the Fraunhofer lines.

1. Discovery of the Fraunhofer Lines and Early Observations of Flame Spectra

Fraunhofer lines and emission spectra later came to be associated with chemistry and physics, but in the beginning the former were discovered and studied together with the latter in relation to certain optical questions. Thus, the English physician, William Hyde Wollaston, who first discovered dark solar lines in 1802, made the discovery while investigating the contemporary question of the number of primary colors to be found in the solar spectrum. Some observers claimed three, while from the appearance of the rainbow others argued for seven. But Wollaston found that if daylight, admitted into a darkened room through a narrow slit, was viewed through a glass prism held close to the eye, only four colors were visible: red, yellowish green, blue, and violet. And the boundaries of these colors were marked by certain dark lines—

The line A that bounds the red side of the spectrum is somewhat confused. . . . The line B, between the red and green, in a certain position of the prism, is perfectly distinct; so also are D and E, two limits of the violet. But C, the limit of the green and blue, is not so clearly marked as the rest; and there are also, on each side of this limit, other distinct dark lines, f and g, either of which, in an imperfect experiment, might be mistaken for the boundaries of these colours.[1]

Wollaston's interest, however, as that of his contemporaries, was in the spectral colors and not in the dark lines, and so the latter were ignored and subsequently forgotten.

Wollaston repeated his experiment using both candlelight and "electric light," and so also became one of the first to observe emission spectra.[2] The blue light of the lower part of a candle flame gave five "images." The first was broad and red and terminated by a bright yellow line, the second and third were green, and the fourth and fifth blue. As for the blue electric light, it also gave several, though somewhat different, images. There Wollaston stopped, thinking it needless minutely to describe appearances which seemed to vary according to the intensity of the light and which he could not anyhow explain. His contemporaries seem to have concurred in this, and we need only note again Wollaston's observation of the yellow line, for it will be met with repeatedly, and will be of much significance in what follows.

Twelve years elapsed before the dark solar lines were independently discovered again in 1814 by Joseph von Fraunhofer, a young German optician employed in an optical-mechanical institute.[3] This time they

[1] W. H. Wollaston, "A method of examining refractive and dispersive powers by prismatic reflection," *Phil. Trans.*, 92 (1802): 365–80; p. 378.
[2] *Ibid.*, p. 380. Flame spectra had been known for at least half a century. See Thomas Melvill, "Observations on light and colours," in *Physical and Literary Essays* (Edinburgh, 1752), 2:35.
[3] Parts of Fraunhofer's papers are to be found translated in J. S. Ames, *Prismatic and Diffraction Spectra, Memoirs by Joseph von Fraunhofer* (New York and London, 1898). This is referred to below as Ames.

were not to be ignored and subsequently forgotten—primarily because of their usefulness in practical optics.

During the course of his work Fraunhofer had sought a means of measuring the dispersive powers of various kinds of glass for light of different colors.[4] For this it was necessary to isolate the latter, yet Fraunhofer did not succeed in finding colored liquids or colored glass which would allow only monochromatic light to pass. And colored flames, he found, did not yield monochromatic spectra corresponding to their color. However, he did find that the spectra of all flames, including oil and tallow flames, possessed in common a sharply defined orange streak, and this "homogeneous" source proved most useful.

Following a series of experiments in which lamplight was used, Fraunhofer commenced yet another series in which he allowed sunlight to pass through a narrow slit into a darkened room and through a prism placed on a theodolite. His own words best describe one of his motives in undertaking this experiment, and also his surprising discovery.

I wished to see if in the colour-image from sunlight there was a bright band similar to that observed in the colour-image of lamplight. But instead of this I saw with the telescope an almost countless number of strong and weak vertical lines, which are, however, darker than the rest of the colour-image; some appeared to be almost perfectly black.[5]

Following a careful investigation, Fraunhofer concluded that the lines were "due to the nature of sunlight" and not to diffraction or illusion. As for the bright orange band of lamplight, Fraunhofer observed that it coincided with the dark line D, as he called it, of the solar spectrum. Later he discovered that in addition to being coincident, both were double.[6]

After Fraunhofer, this bright doublet found in most spectra was described as being yellow. Nevertheless, to Fraunhofer it was decidedly orange, for in the spectrum of the "light of electricity" he observed a line in the orange which at first seemed to have the same color as the bright line of lamplight. Yet, when its angle of refraction was measured it was seen to correspond more with the yellow rays of lamplight.[7]

[4] J. von Fraunhofer, "Bestimmung des Brechungs- und Farbenzerstreuungs-Vermögens verschiedener Glasarten, in Bezug auf die Vervollkommnung achromatischer Fernröhre," *Denkschriften der Königlichen Akademie der Wissenschaften zu München,* 5 (1814–15): 193–226; in Ames, pp. 3–10; p. 3.
[5] *Ibid.,* p. 4.
[6] *Ibid.,* p. 10; and also J. von Fraunhofer, "Kurzer Bericht von den Resultaten neurer Versuche über die Gesetze des Lichtes, und die Theorie derselben," *Ann. der Physik,* 74 (1823): 337–78; in Ames, pp. 41–61; p. 58.
[7] J. von Fraunhofer, "Bestimmung des Brechungs- und Farbenzerstreuungs-Vermögens verschiedener Glasarten"; in Ames; pp. 9, 10.

Fraunhofer's method of producing "electric light" was to place two conductors half an inch apart and join them by means of a glass thread. One conductor was then connected to an electrical machine and the other to earth. While the current was passed, light appeared to stream continuously in a fine luminous line between the poles. A large electrical machine was found to produce many more lines than the "several" observed with a weaker machine.[8]

In all, Fraunhofer found 574 dark solar lines, the strongest of which he mapped. His primary interest in the lines was that they afforded him a means of attaining his original goal of determining refractive indices for most wavelengths. This he could do with great accuracy, as the majority of solar lines are sharply defined. Indeed, their usefulness in this respect, was what attracted attention to the lines, and once they were generally known they naturally demanded an explanation. This question, however, does not seem to have engaged Fraunhofer.

2. Flame Spectra and Chemical Analysis

Fraunhofer's work appeared at a time of intense interest and rapid progress in optics, when the rival wave and particle theories competed for supremacy and when many optical matters considered more important claimed attention. One of the most interesting phenomena of contemporary optics was polarization, discovered in the interval between the publication of Wollaston's and Fraunhofer's papers. This subject was significantly advanced by the Scotsman, David Brewster, who throughout his long and productive life maintained an active interest in most optical matters.

In 1822 Brewster published an account of a lamp designed to produce monochromatic yellow light.[9] This was an improvement upon Fraunhofer's method of producing this light useful in practical optics. However, Brewster's publication is important for another reason, namely, that it was the first of a series of related papers by himself and others, which appeared over the next twenty years and in which an explanation of the Fraunhofer lines was both arrived at and rejected, while a limited form of chemical analysis based on the spectra of colored flames was also attained.

That part of Brewster's work which drew the first response deals

---

8 *Ibid.*, p. 9; J. von Fraunhofer, "Kurzer Bericht von den Resultaten neurer Versuche über die Gesetze des Lichtes;" in Ames, pp. 41–61; p. 58.
9 D. Brewster, "Description of a monochromatic lamp for microscopical purposes, etc., with remarks on the absorption of the prismatic rays by coloured media," *Trans. Roy. Soc. Edinburgh,* 9 (1823): 433–44.

with the absorption of sunlight by transparent colored media. This subject had previously mainly concerned astronomers (such as the celebrated William Herschel) in a practical way, but had also been of some interest to Thomas Young, champion of the wave theory, and, as we recall, to Fraunhofer.[10] Brewster's primary objective was "the discovery of a general principle of chemical analysis, in which simple and compound bodies might be characterized by their action on definite parts of the spectrum."[11] His procedure was to pass sunlight through different colored media and analyze the emergent light by means of a prism. He found, for example, that "blue glass attacks the spectrum in several points at the same time, and after absorbing all the middle rays, it leaves only the extreme red, and the portions of the blue and violet spaces which are contiguous."[12] Brewster did not succeed, however, in arriving at any general principle of chemical analysis.

The response to Brewster's observations came in the same year from John Herschel, then busy with astronomical observations.[13] However, while experimenting several years earlier "on the scale of tints developed by polarized light," Herschel had also used colored media in an attempt to obtain "tolerably homogeneous" rays from sunlight. Finding that the latter was a satisfactory source for all but red and yellow rays he had sought and found these in several sources, including luminous flames. His "few examples of remarkable flames" are of interest, as their spectra are described for the first time.[14] Sulphur, for example, when burning feebly, emits most rays, but principally violet and blue ones; when burning vigorously it emits a perfectly homogeneous and brilliant light, and when mixed with saltpeter and burned this latter definite color gives way to several others.[15]

Three years later, in 1825, when William Henry Fox Talbot, afterward to be associated with the development of photography, wrote on colored flames in the journal of which Brewster was

[10] W. Herschel, "On the power of penetrating into space by telescopes," *Phil. Trans.*, 90 (1800): 49–85; T. Young, "On the theory of light and colours," *Phil. Trans.*, 92 (1802): 12–48.

[11] D. Brewster, "Observations on the lines of the solar spectrum, and on those produced by the earth's atmosphere, and by the action of nitrous acid gas," *Trans. Roy. Soc. Edinburgh*, 12 (1834): 519–30; p. 519.

[12] D. Brewster, "Description of a monochromatic lamp for microscopical purposes." p. 439.

[13] J. F. W. Herschel, "On the absorption of light by coloured media, and on the colours of the prismatic spectrum exhibited by certain flames; with an account of a ready mode of determining the absolute dispersive power of any medium, by direct experiment," *Trans. Roy. Soc. Edinburgh*, 9 (1823): 445–60.

[14] Unfortunately, Herschel's drawings of these spectra were not published.

[15] J. F. W. Herschel, "On the absorption of light by coloured media," p. 455.

editor, these phenomena were "still unexamined, or imperfectly explained."[16] By impregnating the cotton wick of a spirit lamp with common salt Talbot had succeeded in producing a brighter homogeneous light than that given by Brewster's monochromatic lamp. He found it difficult, however, to determine precisely the source of the homogeneous yellow light.

I have found that the same effect takes place whether the wick of the lamp is steeped in the muriate, sulphate or carbonate of *soda*, while the nitrate, chlorate, sulphate, and carbonate of *potash*, agree in giving a blueish tinge to the flame. Hence, the yellow rays may indicate the presence of *soda*, but they, nevertheless, frequently appear where no soda can be supposed to be present.[17]

Talbot had confirmed Herschel's finding that yellow light was given by burning sulphur, a fact which he regarded as indicating "a very singular optical analogy between soda and sulphur, bodies hitherto supposed by chemists to have nothing in common."[18] Talbot also found that candles, and platinum touched by the hand or rubbed with soap, also gave the yellow light. Common salt sprinkled on platinum gave this light while the salt decrepitated, and the effect could be renewed at will by wetting the platinum. This latter circumstance led Talbot to suppose that the light was due to water of crystallization, rather than to sodium, the more so as it was also given by wood, ivory, and paper, whose only common constituent with sodium salts is water. But then there was the problem of explaining why, for example, potassium salts should not also produce the yellow light. Talbot finally concluded that water could not produce it as the light was also produced by sulphur, which was supposed to have no "analogy" with water.[19]

This provides an excellent illustration of the problem of the ubiquitous yellow line, which was to perplex spectroscopists for a considerable time to come and constitute a major obstacle to progress.

Talbot found that potassium salts displayed a characteristic red ray corresponding to the "characteristic" yellow ray of sodium salts, and on this basis he suggested that "whenever the prism shows a *homogeneous* ray of any colour to exist in a flame, this ray indicates the formation or the presence of a *definite chemical compound*."[20] Thus a

---

[16] W. H. F. Talbot, "Some experiments on coloured flames," *Edinburgh J. Sc.,* 5 (1826); 77–81.

[17] *Ibid.,* p. 78.

[18] *Ibid.,* p. 79.

[19] *Ibid.*

[20] *Ibid.,* p. 81.

prismatic analysis of the "red fire of the theatres" would indicate from the bright yellow and orange lines the presence of sulphur (in spite of the yellow rays being characteristic of sodium!) and strontium. And he concluded that: "If this opinion should be correct and applicable to other definite rays, a glance at the prismatic spectrum of a flame may show it to contain substances, which it would otherwise require a laborious chemical analysis to detect."[21]

Talbot's expression is somewhat inconsistent—on the one hand he generalizes in terms of chemical compounds and substances, while on the other he speaks specifically of elements such as sulphur and strontium—yet he clearly expresses the basic idea of chemical spectrum analysis. Indeed, what we have just read finds a striking parallel in what Bunsen and Kirchhoff were to declare much later. Nevertheless, Talbot's analysis was restricted to only a few elements, and was consequently of little practical value. It did, however, mark an advance over the familiar flame analysis in which the salts of sodium, potassium, calcium, strontium, magnesium, lithium, barium, and copper, could be distinguished from one another by the colors which they imparted to flames.[22] In this analysis the presence of one salt could mask the presence of another, but with Talbot's analysis all of the salts present could be identified.

Talbot was aware of the need of extending his analysis, but this was either not easy or people were not sufficiently interested. In 1827 John Herschel described the spectrum of calcium salts as displaying a yellow and a bright green line, but while aware of Talbot's work he did not mention this observation as an addition to the latter's analysis.[23] Much later, Bunsen and Kirchhoff were to distinguish calcium from the alkali metals and from the other alkaline earths by its characteristic green line.[24]

The subject was not further advanced until 1834, some nine years later, when Talbot observed that while ordinarily it was difficult to distinguish the red lithium and red strontium flames, yet by using a prism the most marked distinction "that could be imagined" was apparent.

The strontia flame exhibits a great number of red rays well separated from each other by dark intervals, not to mention an orange, and a very definite

21 Ibid.
22 See J. Herschel's essay on light in Encyclopaedia Metropolitana (London, 1827), p. 438.
23 Ibid.
24 R. Bunsen and G. Kirchhoff, "Chemical analysis by spectrum observations," Phil. Mag., 20 (1860): 89–109.

bright blue ray. The lithia exhibits one single red ray. Hence I hesitate not to say that optical analysis can distinguish the minutest portions of those two substances from each other with as much certainty, if not more, than any other known method.[25]

Nevertheless, this was but a small extension of the analysis.

It is curious that following Talbot's paper of 1825 apparently no one, including Talbot, thought it worth while to carry out a comprehensive and systematic investigation of the flame spectra of salts. This would seem to have been the obvious next step. Eventually, however, it was proposed—by Talbot.

In 1836 he outlined as a desideratum a program similar to that which Bunsen and Kirchhoff later realized together.

. . . an extensive course of experiments should be made on the spectra of chemical flames, accompanied with accurate measurements of the relative position of the bright and dark lines, or *maxima* and *minima* of light which are generally seen in them. The definite rays emitted by certain substances, as, for example, the yellow rays of the salts of soda, possess a fixed and inviolable character, which is analogous in some measures to the fixed proportion in which all bodies combine according to the atomic theory. It may be expected, therefore, that optical researches, carefully conducted, may throw some additional light upon chemistry.[26]

Here we see an enlarged perspective. Previously Talbot had been concerned with the usefulness of flame spectra in practical chemistry, and now he is suggesting their potential importance to theoretical chemistry.

In spite of this change of emphasis it is tempting, on first consideration, to assume that had Talbot carried out his proposed program he would have fully anticipated Bunsen and Kirchhoff, not in the conception—he had already done that—but in the range of application of chemical spectrum analysis. Such an assumption would be quite erroneous, however, for about six years later Brewster partially completed a program similar to that proposed by Talbot, without, however, arriving at any positive results—in regard to either practical or theoretical chemistry.

At the 1842 meeting of the British Association for the Advancement of Science Brewster gave what was intended to be a preliminary report of his researches.[27] Having procured as many minerals, salts, and sub-

25 W. H. F. Talbot, "Facts relating to optical science, No. 2," *Phil. Mag.*, 4 (1834): 112–14; p. 114.

26 W. H. F. Talbot, "Facts relating to optical science, No. 3," *Phil. Mag.*, 9 (1836): 1–4; p. 3.

27 D. Brewster, "On the luminous bands in the spectra of various flames," *Rep. British Assn.*, 11 (1842), pt. 2, pp. 15–16.

stances as possible, he had investigated some "200 or 300." However, not having "grouped" all of the results he merely mentioned a few of them. Of importance is the observation that "Fraunhofer's luminous *D* line" was given by almost every substance. Brewster expressed the hope of being able to present a full report at the following year's meeting of the Association. But he never did.[28]

Chemical spectrum analysis had reached an impasse where it was to remain until the time of Bunsen and Kirchhoff. My research has not directly revealed the reason or reasons for this, but with the benefit of hindsight it may confidently be suggested that perhaps the most important, if not the only, reason was that the apparent complexity of the numerous spectra defied analysis. It has been suggested that the ubiquity of the yellow *D* line was solely responsible.[29] That is certainly part, though not all, of the answer. It is the most notable of many similar complicating factors. Also, the degree of sensitivity of the spectral reaction is greater than had yet been realized, and therefore the consequent necessity of scrupulously avoiding contamination by impurities was not fully appreciated. In addition, as the last quotation from Talbot shows, it was recognized that many flame spectra displayed dark as well as bright lines. Elsewhere Talbot mentions the "remarkable" spectra of copper salts, boric acid, and barium nitrate, which are covered throughout with dark lines "similar to those of the solar spectrum."[30] Here Talbot may have unconsciously been viewing the spectra of undissociated compounds, yet the dark lines still appeared to be important.

In 1848, however, John William Draper, the eloquent Professor of Chemistry at the University of New York, demonstrated that the dark lines of flame spectra were spurious, being due to the presence of some incombustible substance in the flame.[31] He did not elaborate upon this. Yet while Draper reduced the apparent complexity of flame spectra in this respect, he nevertheless added to it in another.

In testing the assertion of "some authors" (and no doubt Draper had Brewster, Herschel, and Talbot in mind) that certain flames give monochromatic light, he examined the flames of many substances and

[28] The British Association apparently considered Brewster's work on absorption more important, for at the 1842 meeting he was requested "to continue his examination of the action of different bodies upon the solar spectrum, with the sum of 40 [pounds sterling] at his disposal for the purpose." *Ibid.,* p. xxii. See also, D. Brewster, "On the action of two blue oils upon light," *Rep. British Assn.,* 12 (1843), pt. 2, p. 8.

[29] See Agnes Clerke, *A Popular History of Astronomy during the Nineteenth Century* (4th ed.; London, 1902), p. 131; and W. A. Hamor, "David Alter and the discovery of chemical analysis," *Isis,* 22 (1934–35): 507–10; p. 508.

[30] W. H. F. Talbot, "Facts relating to optical science, No. 3," p. 3.

[31] W. Draper, "On the production of light by chemical action," *Phil. Mag.,* 32 (1848): 100–14; p. 108.

found that while they were of different colors, yet prismatic analysis showed each to contain "every prismatic colour." For example, the supposedly monochromatic flame of alcohol burned on a wick impregnated with common salt was found to give other, although less strong, colors in addition to yellow.[32] Many flame spectra, therefore, were less simple than had been maintained.

This was a view which became predominant in the 1840s, and which is excellently illustrated by the first published sketches of flame spectra made in 1845 by William Allen Miller, Professor of Chemistry at King's College in London.[33] Miller drew the spectra of calcium, copper and barium chlorides, boric acid, and strontium nitrate, each of which displayed several bright lines and bands, and all having the yellow D line in common. This latter yellow line was the only one visible in the "brilliant" spectra of zinc, iron, steel, and platinum ignited in an oxyhydrogen jet. It was also given by sodium chloride, but Miller noted that the latter, like the first five substances, showed "a marked tendency to the occurence of bands in other parts." The consequence of results like Draper's and Miller's was to obstruct the extension of chemical spectrum analysis, as it became impossible to detect further lines characteristic of only one element.

This growing awareness of the confusing complexity of flame spectra was not unrelated to the now ever-present problem of explaining the origin of the Fraunhofer lines. In 1842 Brewster told the British Association for the Advancement of Science of his surprise in finding with saltpeter that the red ray "occupied the exact place of the line A in Fraunhofer's spectrum," and that a second bright line corresponded to Fraunhofer's line B.[34] In fact, "all the black lines of Fraunhofer were depicted in the spectrum in brilliant red light." He had looked with "great anxiety" to see if there was anything similar in other flame spectra, and it appeared that this was a property of almost all.

The combined effect of these various factors was to make it appear, in the 1840s, that a comprehensive chemical analysis based upon flame spectra was at best unlikely, if not impossible. Thus the earlier hopes of the twenties and thirties were not realized.

3. "Electric Spectra" of Metals

In considering the state of knowledge of the so-called electric spectra of metals during the same period a marked contrast is seen. Although

---

32 *Ibid.*, pp. 101–3.

33 W. A. Miller, "Experiments and observations on some cases of lines in the prismatic spectrum produced by the passage of light through coloured vapours and gases, and from certain coloured flames," *Phil. Mag.*, 27 (1845): 81–91; pp. 89–91.

34 D. Brewster, "On luminous lines in certain flames corresponding to the defective lines in the sun's light," *Rep. British Assn.*, 11 (1842), pt. 2, p. 15.

the subject was studied to an even lesser degree than flame spectra, significant progress was nevertheless made. The principle that each metal has a characteristic spectrum was early established and, as will be seen, this was to be essential to Bunsen and Kirchhoff's achievement. From the early decades of the nineteenth century there was a considerable, though by no means new, interest in the nature of "electric light."[35] No doubt the more powerful voltaic and electromagnetic sources of electricity, just then made available by the science of electricity in one of its most fertile periods, had been instrumental in promoting this interest.[36]

We have noted that in investigating the "spectrum of electric light" Fraunhofer employed a frictional electrical machine. The next important contributor to the subject was Charles Wheatstone, Professor of Physics at King's College in London, who employed both voltaic and electromagnetic sources.[37] Wheatstone gave an account of his experiments and results at the 1834 meeting of the recently formed British Association for the Advancement of Science. From the spectroanalytic standpoint his most important discovery was that the spectrum of electric light varied according to the metallic electrodes employed. The number, positions, and colors of the bright lines varied in each case, and each spectrum was so different from the others that by this means the metals—mercury, zinc, cadmium, tin, bismuth, and lead—could readily be distinguished from one another. If the spark was passed between two different metals the bright lines of each were simultaneously seen.

Significant as this was, however, Wheatstone's interest was primarily in the nature of electric light. He found that when the voltaic spark from mercury was passed successively "in the ordinary vacuum of the air-pump, in the Torricellian vacuum, in carbonic acid gas, etc., . . . the same results were obtained as when the experiment was performed in the air or oxygen gas." And from this incorrect result he correctly inferred that electric light resulted from the "volatilization and ignition (not combustion) of the ponderable matter of the conductor itself."

---

[35] See, for example, C. Despretz, "Sixième communication sur la pile. Note sur le phénomène chimique et sur la lumière de la pile à deux liquides," *Comptes Rendus,* 31 (1850): 418–22; and "De la lumière électrique considérée dans sa plus grande généralité," in A. C. Becquerel, *Traité Experimentale de L'Électricité* (Paris, 1836), 4: 31–38. In the 1855 edition the number of pages given to this subject had been increased to thirty–1:335–66.

[36] For a description and sketch of a popular electromagnetic machine (Pixii's), see A. C. Becquerel, *Traité Experimentale de L'Électricité* (Paris, 1836), 2:205–7.

[37] C. Wheatstone, "On the prismatic decomposition of electric light," *Phil. Mag.,* 7 (1835): 299; and *Rep. British Assn.,* 3 (1834), pt. 2, pp. 11–13. This is a summary of Wheatstone's paper, which was published in full only after Bunsen and Kirchhoff's work appeared. See *Chemical News,* 3 (1861): 198–201.

Curiously, the electric spectra of metals seem to have had little direct appeal at the time. Perhaps it was because his explanation of the origin of electric light was, as he indicates, already generally accepted, that Wheatstone chose not to investigate this subject further. For a considerable time to come the only addition to the subject of electric spectra was made by Talbot, who in 1836 described the "distinctive" spectra of silver, copper, and gold, which he had observed in Faraday's laboratory.[38]

The subject was next advanced about fifteen years later by Antoine Masson, a professor of the Faculty of Sciences at Paris, during the course of a lengthy series of experiments upon electric light.[39] When Masson commenced his experiments in the mid-forties it had already been known for some time that electricity could, by suitable means, be converted into heat, light, or magnetism; and Masson was intrigued by the idea, which he says was shared by others, that electricity, heat, and light, might all be accounted for by the various movements of a single substance. This was at a time when in electricity one had a choice between the one- and two-fluid theories, when the caloric theory of heat was only being challenged by the incipient kinetic theory, and when the wave theory of light had almost displaced its rival, the corpuscular theory. Accordingly, the substance in question was the all-pervading ether. The first step toward such a unifying explanation of all three phenomena was naturally a comprehensive investigation of the relations existing between the three. And as that had already been accomplished to some extent for electricity and heat, Masson set himself the tasks of investigating the production of electric light and of comparing the quantities of heat and light produced by various electric currents.[40]

During the course of his investigations Masson studied electric spectra and confirmed Wheatstone's finding that metals have characteristic spectra. However, he observed that all metallic spectra shared "common lines"—common, that is, in position, but differing in intensity with each metal. Masson gave the first detailed accounts and drawings of the electric spectra of iron, copper, tin, lead, antimony, bismuth, zinc, cadmium, and carbon, in which the common lines—in the red, orange, yellow, and green—were respectively denoted by a, β, γ, and δ.[41]

Although Masson observed the electric spectra of different metals in

[38] W. H. F. Talbot, "Facts relating to optical science, No. 3," pp. 3–4.
[39] A. Masson, "Études de photométrie électrique," *Annales de Chimie et de Physique,* 14 (1845): 129–95; 30 (1850): 5–55; 31 (1851): 295–326; 45 (1855): 385–454.
[40] *Ibid.,* 14 (1845): 132.
[41] *Ibid.,* 31 (1851): 300–1.

various gases, he did not discover—as did Anders Jonas Ångstrom (Professor of Physics at Uppsala) soon afterward, in 1853—that the so-called electric spectrum of a metal was actually an overlapping of two spectra, one corresponding to the metal of the electrode and the other to the particular gas enveloping it.[42] This was a significant discovery for hitherto only absorption spectra of gases had been known, and it had not been suspected that gases could also give emission spectra.[43]

Accordingly, Ångstrom investigated the spectra of air, hydrogen, oxygen, and carbon and nitrogen dioxides. His apparatus consisted of a closed glass tube in which the electric spark passed between two brass electrodes. The gas to be studied was led in at one end of the tube and out at the other. To ensure that the tube contained no atmospheric air, the gas was usually passed through for an hour before observations were begun. It should be particularly noted that the gases remained at atmospheric pressure and that they were not perfectly pure—such was the state of spectroscopy at this juncture that Ångstrom did "not believe that an erroneous result could be produced by the impurity present."[44]

Ångstrom sketched the spectra of the above mentioned gases. His observation that the spectra of oxygen and carbon dioxide might be regarded as identical is of interest because of how he explained the identity: following J. J. Berzelius's teaching the electric spark decomposes the carbon dioxide into carbon monoxide and oxygen, and the latter gas produces its own characteristic lines in the spectrum.[45]

One might reasonably expect that Ångstrom's discovery of gaseous emission spectra would have led to a simplification of electric metallic spectra, in particular to a partial if not complete reduction in the number of Masson's common lines. On the contrary, Ångstrom found all metallic spectra to share in common "a multitude of luminous lines, comparable in number and distribution with the lines of Fraunhofer in the solar spectrum."[46] Among these lines Ångstrom drew attention to two which had first been observed by Fraunhofer and which exceeded the others in brightness, namely, a double line in the yellow and a line in the

[42] A. J. Ångstrom, "Optiska undersökningar," [1853] *Konliga Svenska Vetenskaps Akademiens Handligar,* (1852): 229–323, in *Phil. Mag.,* 9 (1855): 327–42; p. 329.

[43] That gases have characteristic emission spectra was also found two years later by the American physician, David Alter. See "On certain physical properties of the light of the electric spark, within certain gases as seen through a prism," *American J. Sc.,* 19 (1855): 231–14. There is an exceedingly poor article by W. A. Hamor on Alter's contribution to spectroscopy (see footnote 29).

[44] A. J. Ångstrom, "Optiska undersökningar," p. 336.

[45] *Ibid.*

[46] *Ibid.,* p. 329.

green. As Ångstrom pointed out, these are the lines γ and δ of Masson. In addition to this common background of lines, each metallic spectrum also possesses characteristic lines much brighter than the former.[47] These characteristic lines might sometimes include the lines γ and δ, in which case the latter are reinforced and become particularly bright. This is so for the line γ in the bismuth spectrum.

In order to explain the common background of electric spectra, Ångstrom assumed with Faraday and Masson that "the electric spark is produced by a current which propagates itself across, and by means of, ponderable matter, which it heats in the same manner, and according to the same laws, as a voltaic current heats a metallic wire."[48] Thus the common background, or, as Ångstrom regarded it, the electric spark spectrum, is produced by this heated ponderable matter. On the other hand, Ångstrom held that the characteristic lines of each spectrum are due to the vibrations of the metal in the gaseous state.

The next practical advance in the knowledge of electric spectra was one of the events which, perhaps more than any other, helped Bunsen and Kirchhoff to establish spectrum analysis. It will be considered presently in that connection.

4. Attempted Explanations of the Fraunhofer Lines

Meanwhile, the Fraunhofer lines had been receiving considerable attention, without, however, any satisfactory explanation of their origin having been given. A successful explanation was to require the prior establishment of a successful chemical spectrum analysis, in addition to the discovery of a physical relationship between the Fraunhofer lines and terrestrial spectral sources. Each was equally important, and as long as both were unrealized no explanation commanding universal assent was possible. Thus, no one explanation is found which gained in strength or adherents. Instead, several bearing little relation to one another are encountered. These explanations involved the related problem of the explanation of optical absorption, concerning which there was again no consensus; so this was an added complication. It was Kirchhoff's simultaneous solution of these intimately related problems that established a consensus and directed future work.

Here it will be seen what a formidable problem the explanation of the Fraunhofer lines appeared to be, and how progress, when finally begun in the middle-fifties, was quickly cut short.

---

47 *Ibid.*, p. 330.
48 *Ibid.*

In the earliest explanations the Fraunhofer lines were regarded as absorption phenomena—"it is no impossible supposition that the deficient rays in the light of the sun and stars may be absorbed in passing through their own atmospheres."[49] This was John Herschel's view and it was shared by Brewster, who conceived the "original" light of the sun to be continuous from one end of the visible spectrum to the other and the "deficient rays" to be absorbed by the gases generated during the combustion by which the light was produced.[50] This was speculation, however, for in their studies of absorption Brewster and Herschel had between them found only one absorption pattern similar in appearance to the solar spectrum.[51] Knowing this, it is perhaps easier to understand the excitement generated by Brewster's subsequent discovery of an absorption spectrum seemingly (for it involved a curious interpretation) identical with the sun's spectrum.

Brewster had continued with his attempt (mentioned in section 1) to create a method of chemical analysis based on the absorption of portions of the solar spectrum. In this he was partly successful. Then, upon finding that some media absorbed light simultaneously in two or more areas of the spectrum, it occurred to him that the number and intensity of such absorptions might depend upon the number and nature of the elements of which the media are composed. Hence his next step was to investigate the action of "elementary bodies" on the solar spectrum. Beginning with iodine and sulphur vapors he found single effects as anticipated. He next used nitrous acid gas.

The result of this experiment completely destroyed the hypothesis which had appeared so plausible, and presented me with a phenomenon so extraordinary in its aspect . . . extending so widely the resources of the practical optician, and lying so close to the root of atomical science, that I am persuaded it will open up a field of research, which will exhaust the labours of philosophers for centuries to come.[52]

Much to Brewster's surprise lamplight, when passed through a small thickness of the gas, had given a spectrum crossed by hundreds of dark lines and bands, far more distinct than the dark solar lines.

While this, as he says, shattered his hope for a means of chemical analysis, it was most important for his explanation of the Fraunhofer lines. For he had found "a most remarkable coincidence" between the lines of the solar and nitrous acid gas spectra. Or so he thought, for

---

[49] Quoted in D. Brewster, "Report on the recent progress of optics," *Rep. British Assn.*, 1 (1831–32): 308–22; p. 320.
[50] *Ibid.*
[51] D. Brewster, "Observations on the lines of the solar spectrum," p. 525.
[52] *Ibid.*, p. 521.

curiously his way of demonstrating this was not to compare the solar spectrum with the nitrous acid gas spectrum formed by lamplight.

In order to afford ocular demonstration of this fact, I formed the solar and the gaseous spectrum with light passing through the same aperture, so that the lines in one stood opposite those on the other, like the divisions in the vernier and the limb of a circle, and their coincidence or non-coincidence became a matter of simple observation. I then superimposed the two spectra, when they were both formed by solar light, and thus exhibited at once the two series of lines, with all their coincidences, and all their apparent deviations from it.[53]

He then gives the details of these deviations:

. . . there were some strong gaseous lines, and even broad bands, to which I could find no counterpart in Fraunhofer's map of the spectrum. . . . This discrepancy at first embarrassed me, and, as I observed it in parts of the spectrum where Fraunhofer had laid down every line which he had seen with his finest instruments, I abandoned all hope of being able to establish the general principle of their identity.[54]

This is most curious. Brewster had passed sunlight through a gas and had obtained a spectrum which appeared to him to have more lines than the solar spectrum. For some reason he was blind to the fact that the gaseous absorption spectrum produced by sunlight would contain, to begin with, every solar line, and that in order to find the true gaseous absorption spectrum he would have to deduct the solar lines. Those very lines which he considered as frustrating his hope for a general identity, actually formed the nitrous acid gas absorption spectrum. This is what he ought to have compared with the solar spectrum, but instead he was trying to establish the identity of the solar spectrum with a combined solar-gaseous spectrum.

Two courses lay open to Brewster. Either he could accept observation as having shown that no identity existed, or he could consider Fraunhofer's map of the solar spectrum as incomplete and commence the "Herculean task" of making a better one. He would have liked to follow the second course, such was his desire to establish an identity, but he had profound respect for Fraunhofer's work.[55]

This respect was shared by Talbot, with whom Brewster discussed the matter and who subsequently suggested that perhaps a change had occurred in sunlight since Fraunhofer had mapped the solar spectrum. This suggestion was sufficient to set Brewster working upon a new map, and apparently he did find most of the lines which he had sought for

53 *Ibid.*, p. 525.
54 *Ibid.*, p. 526.
55 *Ibid.*

in vain in Fraunhofer's map as the counterpart of those in the gaseous spectrum. Whereas, according to Brewster, Fraunhofer's map contains 354 lines, his contains more than 2,000. He explains that: "The action of the gas upon invisible lines in the spectrum rendered them visible by slightly enlarging them, and this enlargement of a solar line indicated the existence of a corresponding line in the gaseous spectrum."[56] Thus, such is his desire for a complete identity that he is willing to speak of "invisible lines" in his solar map. While to us today the effect of the gas is to produce additional lines in the solar spectrum, to Brewster it was merely to enlarge already existing, although invisible, lines.

In this way the embarrassing discrepancy was removed and Brewster proceeded to draw up three different maps which he apparently considered identical. These were: a map of the solar lines; a map exhibiting the action of nitrous acid gas upon sunlight "previously deprived of a number of its definite rays" (he says no more than this); and a map showing the action of the gas upon a continuous spectrum of artificial white light.[57] We may well wonder why, in the second instance, reduced sunlight was employed. Nevertheless, from the identity of the three maps Brewster was able to declare that the same absorptive elements which exist in nitrous acid gas exist also in the sun's atmosphere, about which John Herschel had earlier said: "we know nothing and can conjecture everything."[58]

Brewster's interpretation was incorrect, however, for his experiments really showed that nitrous acid gas was not present in the sun's atmosphere. Perhaps he allowed himself to be carried away by the thought of being the first to explain the origin of the Fraunhofer lines, and so he saw what he wanted to see and not what actually was. In saying that the effect of the gas was to enlarge already existing but "invisible" lines, he was obscuring a phenomenon which was later to be a decisive tool in Kirchhoff's hands. Again there is the curious fact that he sought a complete, and not a partial, identity; and that when this had been achieved he did not consider it desirable to examine the absorption spectra of other gases to see if the solar atmosphere contained additional elements.

I have dealt with Brewster's work at some length because it was much quoted in contemporary literature. In spite of the contradiction which it contained it was the basis for the interpretation of the Fraunhofer lines as absorption phenomena. In principle this was sound, at least for a few years.

While the Fraunhofer lines might be explained as absorption

---

56 *Ibid.*, p. 527.
57 *Ibid.*, pp. 527–28.
58 J. Herschel, *A Treatise on Astronomy* (Philadelphia, 1834), p. 202.

phenomena, an explanation of the phenomenon of absorption was still wanting. Absorption spectra posed a formidable problem, for they seemed to present no regular relationships and thus little hope of a solution. "We seem," wrote John Herschel, "to lose sight of the great law of continuity and to find ourselves involved among desultory and seemingly capricious relations, quite unlike any which occur in other branches of optical science."[59]

Although Brewster, an ardent supporter of the corpuscular theory, was unable to explain the nitrous acid absorption spectrum in terms of this theory, he did not hesitate to use it in an attack upon the rival wave theory. Writing in 1833 he argues by analogy from another set of wave phenomena.

Among the various phenomena of sound no such analogous fact exists, and we can scarcely conceive an elastic medium so singularly constituted as to exhibit such extraordinary effects. We might readily understand how a medium could transmit sounds of high pitch, and refuse to transmit sounds of a low pitch: but it is incomprehensible how any medium could transmit two sounds of nearly adjacent pitches, and yet obstruct a sound of intermediate pitch.[60]

Indeed the chief objection made to the wave theory, after Cauchy had given his theory of dispersion in 1830, was that it left the phenomenon of absorption unexplained.[61]

In 1833, however, Herschel, following Brewster's analogy and taking the idea of resonance, suggested an explanation of absorption in terms of the wave theory.[62] This was elaborated and expressed more succinctly two years later, in 1835, by Talbot.

We are to assume that light when traversing a transparent medium is able to set the particles of the medium in motion, the particles being disposed to vibrate with a frequency not altogether dissimilar to that of light. If, then, the rays of different color are also of different frequency (as was then beginning to be thought probable) some of them will vibrate *"in accordance"* and others *"in discordance"* with the vibrations of the medium. "And these accordances and discordances will succeed each other in regular order, from the red end of the spectrum to the violet end; each discordance being marked by a line or deficiency in the spectrum, because the corresponding ray is not

59 J. Herschel, "On the absorption of light by coloured media, viewed in connection with the undulatory theory," *Phil. Mag.*, 3 (1833): 401–12; p. 402.
60 D. Brewster, "Observations on the absorption of specific rays, in reference to the undulatory theory of light," *Phil. Mag.*, 2 (1832): 360–63; p. 363.
61 See F. J. von Wrede, "Attempt to explain the absorption of light according to the undulatory theory," *Scientific Memoirs*, ed. Richard Taylor, 1 (1837): 477–82, 483–502; p. 477; and, E. Whittaker, *A History of the Theories of Aether and Electricity* (3rd ed.; New York, 1960), 1: 166.
62 J. Herschel, "On the absorption of light by coloured media," pp. 406 ff.

able to vibrate through the medium, but is arrested by it and absorbed."[63] Thus, according to this view a body transmits all those vibrations which it normally emits, and absorbs all others.

In the same year, however, Brewster's absorption explanation of the origin of the Fraunhofer lines, "that frequent subject of enquiry," was directly tested and invalidated by James D. Forbes, Professor of Natural Philosophy at the University of Edinburgh.[64] Supposing the sun to possess an atmosphere, he reasoned, then it is clear that the absorptive action must be greatest on the light which reaches us from the sun's "edges," and least on that which traverses the atmosphere vertically. Thus light from the "edges" of the solar disc might be expected to display "more numerous and broader" bands than light from its entire surface.[65]

The annular eclipse of 1836 afforded an opportunity of testing this, and careful observation revealed no difference in the number, position, or "thickness" of the Fraunhofer lines. Thus Forbes concluded that the solar atmosphere was not responsible for the production of the lines. Furthermore, he thought that this ought not to be surprising, for occasionally flame spectra also displayed bright and dark bands without there being any reason to suspect absorptive action. Thus, the solar light could also be "primitively incomplete." As is recalled, the spurious nature of the dark lines of flame spectra was not demonstrated until 1848.

Forbes's conclusion appears to have been accepted by the scientific community, and in particular by Brewster, for almost twenty-five years later, and on the eve of Kirchhoff's work, Brewster and J. H. Gladstone verified Forbes's result and stated that the origin of the Fraunhofer lines was still an undecided question.[66]

That is not to say, however, that further attempts had not been made to explain the Fraunhofer lines. But, as W. A. Miller observed, the problem of giving an explanation was one which involved "so much difficulty and obscurity."[67]

One attempted explanation was based on an alternative theory of

63 W. H. F. Talbot, "On the nature of light," *Phil. Mag.*, 7 (1835): 113–18; p. 117.

64 J. Forbes, "Note relative to the supposed origin of the deficient rays in the solar spectrum; being an account of an experiment made at Edinburgh during the annular eclipse of 15th May, 1836," *Phil. Trans.*, 126 (1836): 453–55.

65 "More numerous and broader," probably because Brewster had shown the earth's atmosphere and nitrous acid gas to add some lines and broaden others. It would seem that "breadth" and "thickness" (next paragraph) are loose terms which include "intensity."

66 D. Brewster, and J. H. Gladstone, "On the lines of the solar spectrum," *Phil. Trans.*, 150 (1860): 149–60; p. 159.

67 W. A. Miller, "Experiments and observations on some cases of lines in the prismatic spectrum," p. 81.

absorption advanced by Baron Fabian Jacob von Wrede. In this theory absorption was regarded as an interference phenomenon. Von Wrede supposed the particles of a transparent medium to partially reflect a transmitted beam of light. The reflected beam now travels in the opposite direction, but is soon partially reflected again to travel once more in the original direction. This continues "ad infinitum," giving rise to an endless series of systems of light waves, each of which possesses a feebler intensity than the one which immediately preceded it. If we confine our attention to the first two systems only, their resultant interaction will be determined in the usual manner by the relation between the wavelength and the relative retardation of the two systems. Suppose then that a continuous spectrum of light is passed through a medium which causes a retardation ϕ in one system relative to the other. Then the intensity of every ray whose half-wavelength is ϕ, $\phi/3$, $\phi/5$, $\phi/7$, $\phi/9$, . . . $\phi/(2m-1)$, $\phi/(2m+1)$ will be a minimum; while the intensity of every ray whose half-wavelength is $\phi/2$, $\phi/4$, $\phi/6$, $\phi/8$, . . . $\phi/(2m)$, $\phi/(2m+2)$ will be a maximum. Thus each wavelength whose intensity is a maximum "must appear" absorbed in relation to the others situated between them, and the resultant spectrum must be analogous to that given by iodine or bromine vapor.[68] These latter absorption spectra, similar to that of nitrous acid gas, had been described by W. H. Miller and J. F. Daniell.[69]

Following von Wrede's theory Adolph Erman suggested that the Fraunhofer lines and the nitrous acid gas absorption spectrum were very probably interference phenomena.[70] The retardations necessary for the production of the Fraunhofer lines are supposed to occur not in any atmosphere of the sun but in the earth's atmosphere. This in spite of Brewster having previously shown that the earth's atmosphere was responsible for only a few of the many dark lines.[71] But then it would seem that after Forbes's result the earth's atmosphere was regarded by many as being the major cause of the Fraunhofer lines. This is clearly stated by Ångstrom in his attempted explanation of the latter.

Ångstrom's explanation was similar to Brewster's in that it attempted to explain the solar spectrum in relation to another single spectrum. Ångstrom compared the "electric spectrum" (mentioned in section 3) with the solar spectrum and found that some, but not all, of the bright lines of the former had corresponding lines in the latter. A complete

68 F. J. von Wrede, "Attempt to explain the absorption of light," p. 477.
69 See Report of the April 22, 1833, meeting of the Cambridge Philosophical Society in *Phil. Mag.*, 2 (1833): 381–82.
70 A. Erman, "Sur la loi de l'absorption de la lumière par les vapeurs de l'iode et du brome," *Comptes Rendus*, 19 (1844): 830–37; pp. 831–32.
71 D. Brewster, "Observations on the lines of the solar spectrum," pp. 528–30.

correspondence, he explained, was the less to be expected because the solar lines, "as is generally assumed, are not only due to the action of the [earth's] atmosphere, but are also to be referred to the action of the sun itself."[72] As the dark lines produced in some way by the sun are not known, neither then is the true solar spectrum. Nevertheless, regarding the latter and the "electric spectrum" Ångstrom boldly declared:

The analogy between the two spectra may, however, be more or less complete when abstraction is made from all the minuter details. Regarded as a whole, they produce the impression that one of them is the reversion of the other. I am therefore convinced that the explanation of the dark lines in the solar spectrum embraces that of the luminous lines in the electric spectrum, whether this explanation be based upon the interference of light, or the property of the air to take up only certain series of oscillations. The first view has only one difficulty, and that is to explain how the different retardations, which are necessary for the effect, are produced; this will be the more difficult for the electric spectrum, inasmuch as all these retardations must occur in the inconsiderable mass of air which is in direct contact with the electric current.[73]

This passage is most interesting. First, Ångstrom believed the explanation of the Fraunhofer lines to involve not the characteristic spectral lines of metals, as Kirchhoff would shortly demonstrate, but the so-called common background of electric spectra! Second, Ångstrom does not know what causes the Fraunhofer lines, and while he says that their explanation "embraces" the explanation of emission spectra, yet he is uncertain as to how the latter are formed. As with the explanation of absorption, there are two possibilities—one in terms of interference, the other in terms of vibrations—and he sees a difficulty with the former while saying nothing about the latter. However, elsewhere he explicitly states his preference for the second—the bright lines are produced by vibrating matter.[74] Yet this does not help in showing how the dark lines and bright electric spectral lines are related. Thus the origin of the Fraunhofer lines remained a mystery.

All of Ångstrom's observations which have been mentioned in this and the previous section are found in the one paper. Ångstrom's primary interest was in explaining optical absorption, and he had only turned to a study of electric spectra because he thought it "would not be without interest for the theory of light."[75] In relation to optical absorption, and quite unrelated to the explanation of the Fraunhofer

---

[72] A. J. Ångstrom, "Optiska undersökningar," p. 332.
[73] Ibid.
[74] Ibid., p. 334.
[75] Ibid., p. 329.

lines, he states that a body "when heated so as to become luminous, must emit the precise rays which, at its ordinary temperature, are absorbed."[76] This differs from the Herschel/Talbot view, according to which a body does not absorb those rays which it emits. Ångstrom, however, knew of no experimental demonstration of his proposition. On the other hand, he was aware of the difficulty presented by the fact that the elasticity of a heated body would be different from that of a body at ordinary temperature.

Before leaving Ångstrom it is of interest to note that following the frequently used analogy with sound he believed a medium to absorb not only those vibrations which it most readily assumes, but also those which bear a simple relation to these latter, such as the octave and the third.[77]

A few years later a different and more positive approach to the problem of the Fraunhofer lines was privately considered. Whereas Brewster and Ångstrom sought to explain the solar spectrum in terms of one other spectrum only, the former was now regarded as having a partial correspondence with several terrestrial spectra. Furthermore, the Fraunhofer lines were once again regarded as originating in the solar atmosphere.

Following his fundamental work in thermodynamics, William Thomson, Professor of Natural Philosophy at the University of Glasgow, became interested in the solar lines and wrote in 1854 to his friend George Gabriel Stokes, Lucasian Professor of Mathematics at Cambridge, concerning them.[78] Thomson asked whether any substance besides "sodium" produced bright lines related to the D lines of the solar spectrum, and whether any of the other dark solar lines had corresponding bright lines.[79] In reply Stokes referred to a few coincidences between bright and dark lines noted by Brewster and pointed out that the subject had as yet been barely attacked. Much measurement remained to be completed, but he thought it likely that the results would be very interesting. Stokes did not know of any "pure substance," other than sodium, which gave the bright yellow line. Significantly, he was of the opinion that it "would be extremely difficult to prove, except in the case of gases or substances volatile at a not very high temperature, that the bright line D, if observed in a flame, was not due to soda, such an infinitesimal quantity of soda would be competent to produce it."[80]

76 *Ibid.*
77 *Ibid.*, p. 328.
78 The Stokes/Thomson correspondence on spectroscopy is found in G. G. Stokes, *Mathematical and Physical Papers* (Cambridge, 1904), 4:367–76.
79 *Ibid.*, p. 367.
80 *Ibid.*, p. 368.

Thomson adopted this point of view, and conceiving that each solar line might correspond to a given "substance" he hoped for a qualitative analysis of the sun's atmosphere—if only more coincidences could be established.[81] Previously Stokes had given what he considered might be a plausible physical explanation of the coincidence of the yellow and D lines "by supposing that a certain vibration capable of existing among the ultimate molecules of certain ponderable bodies, and having a certain periodic time belonging to it, might either be excited when the body was in a state of combustion, and thereby give rise to a bright line, or be excited by luminous vibrations of the same period, and thereby give rise to a dark line by absorption."[82] This was similar to Ångstrom's view of absorption, but like Ångstrom, Stokes did not know of a single experiment which justified his view—he was not aware of any absorption bands, identical to the D lines, having been produced by sending light through a vapor.

However, in 1855 Stokes learned of an experimental discovery which did justify his explanation. This discovery had been made and announced six years earlier by Léon Foucault, but had remained relatively unknown. Stokes learned of it directly from Foucault when the two met at dinner while Foucault was in London to receive the Royal Society's Copley medal.

Foucault had become curious as to whether the double yellow line of the electric arc spectrum of carbon corresponded with the double D line of the solar spectrum. Lacking an instrument with which to determine the wavelengths of the two lines, he passed sunlight through the arc itself and viewed the two spectra superposed. He found the lines to be "exactly" coincident. But in addition this arrangement revealed the unexpected and curious fact that the arc absorbed the D lines, so that they became considerably reinforced when the two sets of lines were exactly superposed. Thus Foucault concluded that "the arc offers a medium which emits, on its own account, the D rays, and which, at the same time, absorbs them when these rays come from somewhere else."[83]

Thus in 1855 Stokes had confirmation of his explanation of the coincidence of the dark D and bright yellow lines. Yet Foucault had already known of this for six years and had not been able to make any further progress. This single fact was insufficient. Foucault's experiments with electric spectra had taught him that the yellow line was inherent in all, and this had been confirmed by the researches of Masson

---

[81] *Ibid.*, p. 369.
[82] *Ibid.*, pp. 370–71.
[83] L. Foucault, "Note sur la lumière de l'arc voltaique," *Recueil des Travaux Scientifiques de Léon Foucault* (Paris, 1878), pp. 170–72.

and Ångstrom. He was not prepared, as Stokes was, to suggest that it might be characteristic only of sodium. But even if Stokes had been certain of this fact, he had not the means of developing its consequences. Thus both Stokes and Foucault were in possession of the explanation of the origin of the Fraunhofer lines several years before Kirchhoff, but they were unable to use it in analyzing the solar spectrum because the requisite knowledge of elemental spectra had not yet been won.

5. Bunsen and Kirchhoff and Flame Spectra

In the mid-1850s, then, progress in understanding the Fraunhofer lines seemed just as remote as progress in understanding the flame spectra of chemical substances. On the other hand, considerable progress had been made in understanding the electric spectra of metals, and while this had fruitful, though limited, consequences for the understanding of the Fraunhofer lines, it nevertheless appeared quite unrelated to the understanding of flame spectra. However, this was soon to change and as a result the impediments to an extended spectral analysis of flames were removed, and once this occurred the way was open to a full explanation of the Fraunhofer lines. This section considers how Bunsen and Kirchhoff were enabled to extend the spectral analysis of flames.

In spite of their apparent confusion flame spectra seem in the 1850s to have appealed to several people for various reasons, but only William Swan, a professor at the Scottish Naval and Military Academy in Edinburgh, found courage to investigate the subject.[84] He undertook to do what Thomson had vainly tried to persuade Stokes to do, namely, to accurately determine the positions of the bright lines of spectra.[85] Whatever Swan's ultimate hope was, his immediate one in 1857 was the explanation of "the phenomena of artificial light," that is, light other than sunlight. So he commenced by examining the flames of burning hydrocarbons.

It is not, however, the results of these experiments which are important here, but rather an unexpected discovery which Swan made during the course of the experiments. In these he considered it imperative to use a colorless flame, and a means of producing such had recently been devised by Bunsen and his student H. E. Roscoe, in their investigations

[84] W. Swan, "On the prismatic spectra of the flames of compounds of carbon and hydrogen," *Trans. Roy. Soc. Edinburgh*, 21 (1857): 411–30.

[85] At the time Thomson himself was engaged in an intensive experimental investigation of certain aspects of thermoelectricity. See Bernard S. Finn, "Thomson's dilemma," *Physics Today*, 20 (1967), no. 9, pp. 54–59; p. 56.

of chemical photometry.[86] Swan's use of Bunsen's flame is an instance of a curious interaction, for the flame allowed Swan to make his discovery, which in turn provided Bunsen and Kirchhoff with one of the clues to an extended spectrum analysis.

Swan describes the flame of the Bunsen lamp as consisting of "at least" two distinct portions—an envelope of a pale lavender tint enclosing a luminous hollow cone of a strong bluish-green color. He noticed that because of its inherent luminosity this outer envelope was highly susceptible to having its color influenced by foreign matter. In dusty air, for example, the flame displayed an abundance of yellow scintillations. His curiosity thus aroused, Swan resolved to determine "how small a portion of matter would in this way render its presence sensible." He chose to experiment with common salt, as the salts of sodium were "well known to be remarkably energetic in producing homogeneous yellow light."[87]

As Swan's work constitutes the turning point in our story and as he expresses himself lucidly and economically, I give in full his description of the experiment and his conclusion.

One-tenth of a grain of common salt, carefully weighed in a balance indicating 1/100 of a grain, was dissolved in 5000 grains of distilled water. Two perfectly similar slips of platinum foil were then carefully ignited by the Bunsen lamp, until they nearly ceased to tinge the flame with yellow light; for to obtain the total absence of yellow light is apparently impossible. One of the slips was dipped into the solution of salt, and the other into distilled water, the quantity of the solution of salt adhering to the slip, being considerably less than 1/20 grain, and both slips were held over the lamp until the water had evaporated. They were then simultaneously introduced into opposite sides of the flame; when the slip which had been dipped into the solution of salt, invariably communicated to a considerable portion of the flame a bright yellow light, easily distinguishable from that caused by the slip which had been dipped into pure water. It is thus proved that a portion of chloride of sodium, weighing less than 1/1,000,000 of a grain is able to tinge a flame with bright yellow light; and as the equivalent weights of sodium and chlorine are 23 and 35.5, it follows, that a quantity of sodium not exceeding 1/2,500,000 of a troy grain renders its presence in a flame sensible. If it were possible to obtain a flame free of yellow light, independently of that caused by the salt introduced in the experiment, it is obvious that a greatly more minute portion of sodium could be shown to alter appreciably the colour of the flame. It therefore follows, that much caution is necessary in referring the phenomena of the spectrum of a flame to the chemical constitution of the body undergoing combustion. For the brightest line in the spectrum of

---

86 R. Bunsen and H. E. Roscoe, "Photo chemical researches—Part 1. Measurement of the chemical action of light," *Phil. Trans.*, 147 (1858): 355–80; pp. 377–79.
87 W. Swan, "On the prismatic spectra of the flames of compounds of carbon and hydrogen," p. 413.

the flame of a candle,—the yellow line R of Fraunhofer,—can be produced in great brilliancy, by placing an excessively small portion of salt in a flame, in whose spectrum that line is faint or altogether absent. The question then arises, whether this line in the candle flame is due to the combustion of the carbon and hydrogen of which tallow is chiefly composed, or is caused by the minute traces of chloride of sodium contained in most animal matter. When indeed we consider the almost universal diffusion of the salts of sodium, and the remarkable energy with which they produce yellow light, it seems highly probable that the yellow line R, which appears in the spectra of almost all flames, is in every case due to the presence of minute quantities of sodium.[88]

The importance of this discovery is twofold. The first is obvious—it led immediately to a simplification of many spectra, including all electric spectra of metals. Perhaps more important, however, is the fact that it underlined the unprecedented sensitivity of reaction met with in spectroscopy. If any certain knowledge was to be gained one was required to operate in conditions of the highest purity. One could no longer afford to be careless of impurities as Ångstrom had been.

Swan's discovery must surely have posed the question: If the ubiquitous yellow line is due to the presence of sodium, do not then other lines common to several electric and flame spectra correspond to but one substance or metal? Suppose one was to spectroscopically examine *highly purified* salts of other metals, would one then find a few characteristic lines associated with each metal? Two years later similar questions determined the course of Bunsen and Kirchhoff's researches. But unfortunately one cannot tell from published sources how they came to conceive their program. However, a discovery by another investigator probably played a vital part in its formulation.

In April of 1859, V. S. M. van der Willigen, a teacher of mathematics in Deventer, Holland, published an abstract of some earlier papers in Poggendorf's *Annalen*.[89] These testify to the continuing interest in the nature of electric light, an interest which was brought into sharper focus in 1857, when the Dutch Society of Science set two prize questions demanding full knowledge of the essential nature of electric light.[90] In investigating electric spectra van der Willigen employed as pole pieces, metallic wires of one millimeter in diameter. When these were moistened with water the spectrum showed a momentary but very strong yellow portion which van der Willigen attributed, on the basis of Swan's work, to sodium impurities. Experimenting in a similar way with hydrochloric acid he found that the effect of the acid was to

88 *Ibid.*, pp. 413–14.
89 V. S. M. van der Willigen, "Ueber das electrische Spectrum," *Ann. der Physik.*, 106 (1859): 610–32.
90 *Ibid.*, p. 610.

intensify the metallic spectral lines. Van der Willigen was able to offer an explanation of this from later experiments in which he found that the metallic lines were similarly intensified when the metals were burned in chlorine. The metallic chlorides produced, being volatile, allowed the metallic parts to become widely dispersed, and in this state they appeared "especially capable" of producing the characteristic bands.[91]

Following this explanation, or in extending the experiments with water and acids, van der Willigen moistened platinum (as this metal gave no "noticeable characteristic bands") pole pieces with weak solutions of "calcium chloride, barium chloride, strontium chloride, calcium nitrate, etc., and always obtained other metallic bands, which are unquestionably characteristic of the metal."[92] The chlorides were particularly effective and calcium nitrate was "serviceable," but the sulphates he found did not give the effect at all. He concluded that: "By this means a wholly new field was opened for my investigations, whereby I could bring all metals without distinction into the spectrum; however my investigations which this part touches are still not finished."[93] As it stood this afforded a means of chemical analysis, but one almost as severely limited as Talbot's and capable of distinguishing merely one chloride, and possibly one nitrate, from another. But the important point is that, because of the previously established principle of the characteristic electric spectra of metals, van der Willigen was able to say something meaningful about the spectra of certain chemical compounds —metal chlorides give spectra identical to those of the component metals. The triumphant achievement of Bunsen and Kirchhoff was to extend this discovery to include the unresponsive sulphates and other salts, so that by placing practically any salt in a flame its component metal could be immediately determined.

The apparatus used by Bunsen and Kirchhoff was much more refined than any hitherto employed. It had a collimator as previously devised by Swan and the prism was housed in a blackened box, no doubt designed to cut out extraneous light.[94] Bunsen's skill and knowledge as a chemist ensured that Swan's admonition concerning pure sources was scrupulously observed. The purification of the salts was carried out as much as possible in platinum vessels and as many as fourteen precipitations or crystallizations were performed. The salt to be exam-

91 *Ibid.*, p. 617.
92 *Ibid.*, pp. 617–18.
93 *Ibid.*, p. 618.
94 See W. Swan, "Note on Professors Kirchhoff and Bunsen's paper 'On chemical analysis by spectrum observations,' " *Phil. Mag.*, 20 (1860): 173–75; p. 175.

ined was melted on to the eye of a suitably bent platinum (following van der Willigen) wire and held in a colorless flame before the collimator.

By this means Bunsen and Kirchoff systematically examined the chlorides, bromides, iodides, hydrated oxides, sulphates, and carbonates of potassium, sodium, lithium, strontium, calcium, and barium. And as a result of these "somewhat lengthy" investigations they concluded that:

The different bodies with which the metals employed were combined, the variety of the chemical processes occurring in several, and the wide differences of temperature which these flames exhibit, *produce no effect upon the position of the bright lines in the spectrum which are characteristic of each metal.*[95]

This was an enormous simplification—instead of there being a different spectrum for each salt there are only as many spectra as there are metals. In other words, a metal gives the same characteristic spectrum no matter whether, or how, it is combined chemically. An accompanying colored plate gave the spectra of the chlorides.

It is significant that in order to demonstrate still more conclusively that each of the above-mentioned metals always produces the same bright spectral lines, Bunsen and Kirchhoff compared the flame spectra of the chlorides with their electric spectra. They produced the latter by sending the discharge of a Ruhmkorff coil through a glass tube containing two platinum electrodes, to each of which small pieces of the selected metal were fastened.[96]

Thus, the chemist was provided with a much more delicate means of chemical analysis than he had hitherto commanded. Nevertheless, the Bunsen/Kirchhoff work had its restrictions. It had merely extended the principle of characteristic spectra, already established for metals and gases, to include the salts of certain metals—the alkalis and the alkaline earths. It did not say anything about the spectra of chemical compounds which were not salts, and indeed it did not reveal all that was to be known about the spectra of salts. This we shall see in dealing with the spectra of undissociated chemical compounds in the next chapter.

Furthermore, the spectra described by Bunsen and Kirchhoff were over-simplified. There are two reasons for this. The limitations of their apparatus allowed them to see only the more conspicuous character-

[95] R. Bunsen and G. Kirchhoff, "Chemical analysis by spectrum observations," pp. 91–92. See also, R. Bunsen, "Ueber Benutzung der Flammenspektren bei der chemischen Analyse," *Verhandlungen des Naturhistorisch-Medicinischen Vereins zu Heidelberg* (1859–60), pp. 31–32.
[96] R. Bunsen and G. Kirchhoff, *ibid.*, p. 93.

istics of a spectrum. But then for the purpose of chemical analysis that was all that was necessary. The chemist needed only to know which line, say, was given by sodium and not by potassium, or by lithium and not by strontium, in order to distinguish his salts. But this is not to say that the Bunsen/Kirchhoff spectra were any less accurate than, say, those of Draper and Miller, with which they stand in great contrast. In fact they were more accurate, certainly more useful and much less complex. Only in relation to what they would afterward be, were they over-simplified. This over-simplification was necessary for the removal of a complex situation, yet it was but a prelude to another complex, though more manageable, situation.

6. Kirchhoff's Explanation of the Fraunhofer Lines

It was after Bunsen and Kirchhoff had begun their collaboration, and before their first joint paper on chemical analysis had been published, that Kirchoff came upon the explanation of the Fraunhofer lines by himself. In addition to his highly successful electrical interests, Kirchhoff was interested in the continuing problem of the absorption of light by colored media, for only one or two years earlier he had assisted a student of Bunsen's, R. Cartmell, who was attempting at Bunsen's suggestion to devise a means of chemical analysis based on absorption.[97] As we recall this was not a new idea, but its recurrence is important for it may have been what led to the Bunsen and Kirchhoff collaboration.

To disgress briefly, Cartmell, whose work was published in 1858, tried to devise a means of detecting the nonvolatile alkalis in one another's presence. He knew that the yellow sodium light obscured that of the other alkalis, but if it could be removed through absorption by a colored medium, then the presence of the other alkalis could possibly be detected from the transmitted light. Cartmell met with limited success but the method was cumbersome. That Bunsen had such research carried out tells us that in 1858 he had as yet not arrived at the idea of a chemical spectrum analysis, for how much easier it was soon to be to heat the mixture of alkalis in a flame and examine the emitted light spectroscopically to tell immediately which alkalis were present.

It is recalled that the coincidence of the bright yellow line, now known to belong to sodium, with the dark *D* line of the solar spectrum had probably been the most intriguing fact of spectroscopy. It con-

---

[97] R. Cartmell, "On a photochemical method of recognizing the non-volatile alkalies and alkaline earths," *Phil. Mag.*, 16 (1858): 328–33. Kirchhoff says Bunsen suggested this research: *Researches on the Solar Spectrum and the Spectra of the Chemical Elements*, trans. H. E. Roscoe (London, 1862–63), 1: 6.

tinued to be so, and in 1857 Swan considered the "question of so much interest and importance," whether:

from the well known coincidence discovered by Fraunhofer, to exist between the line R, in the spectrum of a lamp, and D of the solar spectrum, taken in connection with similar phenomena, which have since been observed,[98] it might be inferred, as a general law of the spectra of flames, that their bright lines always coincide with dark lines of the solar spectrum.[99]

Swan accordingly compared the bright lines of the carbohydrogen spectrum[100] with the dark solar lines and found that for the most part the former coincided, not with dark lines but with bright spaces of the solar spectrum. While this was a disappointment he did not lose all hope, and observed that in cases

where there is a remarkable analogous configuration of the two groups of lines, accompanied by exact coincidence, as between the double lines *"a"* and D; and more especially where we actually view the striking phenomenon of the lines superimposed, the impression of some physical connection between the two groups becomes irresistible.[101]

In reading these views of Swan, Kirchhoff would also have learned of Foucault's work mentioned above, for Swan refers to it in a footnote. He refers to it, to be sure, not as an instance of the physical relationship which he was seeking, but as supporting evidence for the coincidence of the yellow and D lines. Thus it is possible that Kirchhoff knew of it only in this latter sense. Kirchhoff, however, later denied knowing about Foucault's discovery and stated that he had independently discovered the same effect.[102] But matters of this nature are of little importance. What is important is that Kirchhoff was able to proceed further than Stokes and Foucault, because unlike them he was now certain that the yellow line belonged to sodium. And in addition he knew the characteristic lines of the flame spectra of other metals from his work with Bunsen (not yet published).

Kirchhoff viewed the spectrum of sunlight passed through a "powerful" salt-flame placed before the slit of his apparatus.[103] This was the

[98] Swan here refers to the coincidences noted much earlier by Brewster.
[99] W. Swan, "On the prismatic spectra of the flames of compounds of carbon and hydrogen," p. 420.
[100] From experiment Swan had found that "probably all substances of the form C_rH_s and $C_rH_sO_t$ produce, when burning, spectra which are absolutely identical." *Ibid.*, p. 419.
[101] *Ibid.*, p. 426.
[102] G. Kirchhoff, *Researches on the Solar Spectrum*, 1: 13.
[103] G. Kirchhoff, "Ueber die Fraunhofer'schen Linien," *Monatsberichte Akad. Wissen. Berlin* (1859): 662–65; in G. G. Stokes, "On the simultaneous emission and absorp-

method he and Bunsen had adopted in comparing flame spectra with the solar spectrum. If the sunlight was sufficiently weak two bright lines appeared in place of the two dark D lines. On the other hand, if the sun's intensity exceeded a certain limit, the two D lines appeared more distinct than if the salt flame had not been employed. This latter surprised Kirchhoff as being totally unexpected. Even more striking he found that the lithium flame, whose bright line had no corresponding dark line in the solar spectrum, could produce a dark line in the latter when sunlight was passed through the flame. There was no published optical theory which could explain these instances, and particularly the second in which two superimposed bright sources produced a dark line! Interference was certainly out of the question as coherent sources were required for that.

Only gradually did Kirchhoff come to their explanation, and his published papers indicate the not always certain route by which he passed from the problem to its solution. By the time of publication of his first paper he and Bunsen had already discovered a clue, namely, that for a dark line to be produced in the spectrum of one emitting source by passing its light through a second emitting source, the second source has to be at a lower temperature than the first.[104]

This fact orientated Kirchhoff's thought toward thermal radiation, which physicists of the 1850s were regarding with more and more certainty as being identical with light, the two being regarded as different "qualities" of the same entity.[105] Kirchhoff's ultimate solution, accounting as it did for both light and thermal radiation, provided further evidence for the essential identity of the two. Furthermore, as an instance of simultaneous discovery Kirchhoff's principle of absorption, more or less, was previously arrived at by Balfour Stewart.[106] However, Stewart was working in the mainstream of thermal radiation studies and was not concerned with explaining the Fraunhofer lines.

As the starting point of his proposed explanation Kirchhoff, as Stewart, takes the law experimentally demonstrated earlier by, among others,

tion of rays of the same definite refrangibility; being a translation of a portion of a paper by M. Léon Foucault, and of a paper by Professor Kirchhoff," *Phil. Mag.* 19 (1860): 193–97; p. 195. See also G. Kirchhoff, "Ueber das Sonnenspectrum," *Verhandlungen des naturhistorisch-medicinschen Vereins zu Heidelberg* (1857–59): 251–55.

104 In G. G. Stokes, *ibid.*, p. 196.

105 See, for example, Baden Powell, "On the analogies of light and heat," *Proc. Royal Institution*, 1 (1851–54): 172–78.

106 B. Stewart, "An account of some experiments on radiant heat, involving an extension of Prevost's law of exchanges," *Rep. British Assn.*, 27 (1858), pt. 2, pp. 23–24; and *Trans. Roy. Soc. Edinburgh*, 22 (1861): 1–20.

F. de La Prevost and P. Desdains, namely, that at the same temperature the ratio between the thermal radiating and absorbing powers of all bodies is the same.[107] In this it was tacitly assumed that all the rays were homogeneous, but in the 1850s it was common knowledge that the radiation emitted by a heated body was heterogeneous. Thus Kirchhoff proceeds to show theoretically that the same law holds for each ray, so that for any ray the ratio of the radiating and absorbing powers of all bodies at a given temperature is the same. And the term "ray" refers both to thermal and to luminous radiation, Kirchhoff assuming that all absorbed rays are converted into heat.

From this Kirchhoff infers that if the spectrum of a red-hot body shows discontinuities or strongly marked maxima or minima, then the absorptive power of that body, regarded as a function of wavelength, must display similar discontinuities or strongly marked maxima or minima. Thus, to consider an example from spectroscopy, the spectrum of the lithium chloride flame consists of "a single bright red line of light":

for waves of this length the radiating power of the flame is very considerable, while for waves of lengths corresponding to the other visible colours it is imperceptible. Accordingly, the power of absorption of the lithium flame must be great for waves of this length, but very small for those constituting the other visible rays. If, therefore, a continuous spectrum be formed by suitable means and a lithium-flame be placed between the source of light and the slit of the apparatus, the spectrum is only affected in the place of the lithium line, its brightness being increased in that part by the radiation of the flame, while on the other hand it is diminished by its power of absorption for waves of that particular length.[108]

This is fine, but Kirchhoff still has to explain why the flame apparently absorbs more light than it emits, giving rise to a dark line. Here he is on uncertain ground. He could not fully explain the effect.

Suppose the absorptive power of the lithium flame to be one-quarter, says Kirchhoff. This flame would then be without effect on the spectrum produced by any other source of light, provided the intensity of the latter were also one-quarter. "If the source of light were proportionately brighter than this, the joint effect of it and the lithium-flame would be to produce a comparatively dull line on a bright ground; and conversely, if the source of light were proportionately duller, a bright line on a dull ground would become visible."[109] However, this is no explanation.

---

[107] G. Kirchhoff, "On the relations between the radiating and absorbing powers of different bodies for light and heat," *Phil. Mag.*, 20 (1860): 1–21.
[108] *Ibid.*, p. 14.
[109] *Ibid.*

When Kirchhoff next attempted to explain "reversals" he was much more lucid, although his explanation was still incomplete. The sodium lines are this time taken as an example, and without referring to his theoretical paper Kirchhoff states that the reversal may be "easily explained upon the supposition that the sodium flame absorbs rays of the same degree of refrangibility as those it emits, whilst it is perfectly transparent for all other rays."[110] If one considers then the effect of a sodium flame upon the light emitted by an incandescent platinum wire, the intensity of the light is altered in the position of the sodium lines. This is for the same reasons as before: the light emitted by the platinum wire is reduced in intensity by a certain amount due to absorption in the flame, while the light emitted by the flame itself is added to it. Thus returned to the crucial point of explanation Kirchhoff continues:

> It is plain that if the platinum wire emits a sufficient amount of light, the loss of light occasioned by absorption in the flame must be greater than the gain of light from the luminosity of the flame; the sodium lines must then appear darker than the surrounding parts, and by contrast with the neighbouring parts they may seem to be quite black, although their degree of luminosity is necessarily greater than that which the sodium flame alone would have produced.[111]

Certainly it is plain that it is so—for that is what Kirchhoff has experimentally observed, but it is not plain that it should be so, that is, that the flame should apparently absorb more radiation than it itself emits.

We might also feel that Kirchhoff was further obliged to explain why the absorbed radiation was not all re-emitted in the original direction, but then Kirchhoff held that upon absorption all radiation was converted to heat.

Thus while Kirchhoff did not fully explain the origin of the Fraunhofer lines, he was able in several instances to demonstrate experimentally their physical connnection with the spectra of metals. But the lack of a complete explanation was no barrier to the task of determining the composition of the sun's atmosphere, which was now commenced by Kirchhoff and others.

So in a sense Kirchhoff came no nearer to a complete explanation of the Fraunhofer lines than did Stokes. But Kirchhoff was able in several cases to obtain new, or accentuate existing, reversals. His was essentially an experimental achievement, and one to which the prior establishment of the principle of chemical spectrum analysis was indispensable. What Bunsen and Kirchhoff did was to lay the experimental founda-

---

[110] G. Kirchhoff, *Researches on the Solar Spectrum*, 1:14.
[111] *Ibid.*

tions of spectrum analysis.[112] A multitude of theoretical questions remained for posterity's consideration.

One final word. The nineteenth-century German historian of physics, F. Rosenberger, states that the essence of Bunsen and Kirchhoff's achievement was theoretical rather than experimental. As opposed to providing decisive new facts, he says, they recognized for the first time in the existing facts the principle of spectrum analysis.[113] As we have seen, however, several persons had conceived the principle of spectrum analysis, but only Bunsen and Kirchhoff were able to provide the facts necessary for spectrum analysis to be effectively established.[114]

112 For contemporary nineteenth-century accounts of the history of spectrum analysis see: W. A. Miller, "On spectrum analysis," *Chemical News,* 5 (1862): 201–3, 214–18; G. Kirchhoff, "Contributions towards the history of spectrum analysis and of the analysis of the solar atmosphere," *Phil. Mag.,* 25 (1863): 250–62; and W. H. F. Talbot, "Note on the early history of spectrum analysis," *Proc. Roy. Soc. Edinburgh,* 7 (1869–72): 461–66.
113 F. Rosenberger, *Die Geschichte der Physik* (Braunschweig, 1887–90), 2:691.
114 This plays havoc with B. M. Kedrov's philosophy of science, for he uses Rosenberger's view of the Bunsen/Kirchhoff work to support his argument that all important discoveries are theoretical generalizations. Also, Kedrov is wrong about the timing of the various parts of the Bunsen/Kirchhoff work. See "Toward a methodological analysis of scientific discoveries," *Soviet Studies in Philosophy,* 1 (1962–63), no. 1, pp. 45–57.

II

||||||||||||||||||||||||||||||

Atoms and Molecules and the Further Extension of the Principle of Spectrum Analysis

D URING THE DECADE of the 1860s spectroscopy was of interest primarily to chemists and astronomers. The former were concerned
with extending the scope of spectroscopy as a tool of chemical
analysis, by which they might not only distinguish between known
elements and their compounds, both inorganic and organic, but also
discover new elements. Astronomers, on the other hand, were interested in employing the chemists' findings in an attempt to determine
the chemical composition of the sun and stars.

In the pursuit of these interests yet others suggested themselves, and
it was in this way that physicists became increasingly interested in
spectroscopy. The further extension of the principle of spectrum analysis, through the discovery first of compound and then of multiple
spectra, is considered in sections 2, 3, and 4. The extension to multiple
spectra involved fundamental atomic and molecular questions and
produced a significant change in the physical view of matter. In order
to appreciate this it is necessary to begin in section 1 by considering
chemical and particularly physical views of atoms and molecules in the
1860s and early 1870s, before an interpretation of multiple spectra was
given. Specifically, the molecular theory of spectra is discussed. In the
final section Lockyer's hypothesis of the dissociation of the elements,
which is closely related to the earlier considerations of the chapter,
is considered.

1. The Molecular Theory of Spectra

Spectroscopy got under way about the same time as, and evolved
side by side with, the kinetic theory of gases; and as each deals with

the particles of gases it was naturally desirable that their teachings on the nature of these particles should be in harmony. Consequently, in order to appreciate the physicist-spectroscopist's understanding of the production of spectra from the 1860s, it is necessary to begin by considering what kinetic theorists were already saying about the nature of gaseous particles. Thus, as it turns out, this first section is largely an account of James Clerk Maxwell's evolving views on the subject.

The various gases such as hydrogen, oxygen, nitrogen, and their several compounds became recognized only toward the end of the eighteenth century. Through the early nineteenth century gases were of interest mainly to chemists, but toward the middle of the century the physical properties of gases began also to be investigated. This latter development coincided with the decline of the caloric theory of heat and the reinstatement, in its place, of the kinetic view of heat. Then in 1857, Rudolph Clausius effectively initiated the kinetic theory of gases with a paper entitled "The Nature of the Motion which we call Heat."

At one point in this paper Clausius argues that in addition to translational motion the "molecules" of a gas may also possess rotational motion. He is also of the opinion that vibrations occur within the moving molecules. "Such vibrations are conceivable in several ways. Even if we limit ourselves to the consideration of the atomic masses solely, and regard these as absolutely rigid, it is still possible that a molecule, which consists of several atoms, may not also constitute an absolutely rigid mass, but that within it the several atoms are to a certain extent movable, and thus capable of oscillating with respect to each other."[1] Clausius adds that in allowing the atoms themselves to move, he does not exclude the hypothesis that each atom may have a quantity of "finer matter," which, without separating from the atom, may nevertheless be capable of movement. This "finer matter" is apparently a quantity of luminiferous ether. We shall see that several of Clausius' fellow countrymen were later to adopt a similar view of the atom.

Clausius found for a volume of gas that the ratio of K, the *vis viva* of the translatory motion of the molecules, to H, the total *vis viva* or heat of the gas, is a constant:

$$\frac{K}{H} = \frac{3}{2} \frac{\gamma' - \gamma}{\gamma}$$

where γ and γ' are the specific heats at constant volume and constant pressure respectively.

[1] R. Clausius, "The nature of the motion we call heat," *Phil. Mag.*, 14 (1857): 108–27; in Stephen G. Brush, *Kinetic Theory* (London, 1966), 1:111–34; p. 113.

Clausius' atomic/molecular view of matter is essentially a chemical one. He speaks, for example, of a molecule of hydrogen chloride being composed of an atom of hydrogen compounded with an atom of chlorine. In addition he assumes "that the force which determines chemical combination, and which probably consists of a kind of polarity of the atoms, is already active in simple [elementary] substances, and that *in these likewise two or more atoms are combined to form one molecule.*"[2]

In his second paper on kinetic theory, published in 1859, Clausius introduced the concept of mean free path, l, given by the equation

$$\frac{1}{l} = \frac{4}{3} \pi S^2 N$$

where S is the radius of the "sphere of action" of a molecule, and N the number of molecules per unit volume.[3] It was this paper which stimulated James Clerk Maxwell's active interest in the subject. In describing his initial investigations to George Gabriel Stokes in May 1859, Maxwell writes:

As we know nothing about either "S" or "N," I thought that it might be worthwhile examining the hypothesis of free particles acting by impact and comparing it with phenomena which seem to depend on this "mean path." I have, therefore, begun at the beginning and drawn up the theory of the motions and collisions of free particles acting only by impact, applying it to internal friction of gases, diffusion of gases, and conduction of heat through a gas (without radiation).[4]

Maxwell gives the results he obtained, adding that he has been unable to check them against experiment. He then asks Stokes if he knows of a refutation of this theory of gases such "as would make it absurd to investigate it further so as to found arguments upon measurements of strictly 'molecular' quantities before we know whether there be any molecules?"[5] Thus Maxwell is not at all certain that molecules actually exist. He carefully avoids using the word "molecule," preferring instead the less definite term "particle."

Maxwell read his paper on kinetic theory at the meeting of the British Association for the Advancement of Science in September of 1859. Then in October he again wrote to Stokes saying that he intended to arrange

2 *Ibid.*, p. 124.
3 R. Clausius, "On the mean lengths of the paths described by the separate molecules of gaseous bodies," *Phil. Mag.*, 17 (1859): 81–91; in Brush, *Kinetic Theory,* 1:135–37.
4 Maxwell's letter is published in *Memoir and Scientific Correspondence of Sir George Gabriel Stokes* (Cambridge, 1907), 2:8–11; p. 8.
5 *Ibid.*, p. 9.

his "propositions about the motions of elastic spheres in a manner independent of the speculations about gases."[6] The paper duly appeared in the *Philosophical Magazine* and its stated purpose was to demonstrate the laws of motion of "an indefinite number of small, hard, and perfectly elastic spheres acting on one another during impact."[7] However, Maxwell also remarks that:

Instead of saying that the particles are hard, spherical, and elastic, we may if we please say that the particles are centres of force, of which the action is insensible except at a certain small distance, when it suddenly appears as a repulsive force of very great intensity. It is evident that either assumption will lead to the same results.[8]

Thus he is also prepared to allow a Boscovichean-type particle. Throughout the paper he repeatedly uses the terms "minute parts" and "particles" of a gas.

There is another contrast with Clausius in that Maxwell does not assume vibrational motions for the particles, elastic spheres, or centers of force of the gas:

If we suppose those aggregate molecules which move together to have a bounding surface which is not spherical, then the rotatory motion of the system will store up a certain proportion of the whole *vis viva,* as has been shown by Clausius, and in this way we may account for the value of the specific heat being greater than on the more simple hypothesis.[9]

Maxwell adds the further refinement that the final state of "any number of systems of moving particles of any form is that in which the average *vis viva* of translation along each of the three axes is the same in all the systems, and equal to the average *vis viva* of rotation about each of the three principal axes of each particle."[10] This is the first expression of the equipartition principle. Maxwell goes on to say that adding "the *vires vivae* with respect to the other axes, we find that the whole *vis viva* of translation is equal to that of rotation in each system of particles, and is also the same for different systems."[11]

However, this result:

(which is true however nearly the bodies approach the spherical form, provided the motion of rotation is at all affected by the collisions) seems decisive against

6 *Ibid.,* 11–14; p. 11.
7 J. C. Maxwell, "Illustrations of the dynamical theory of gases," *Phil. Mag.,* 19 (1860): 19–32 and 20 (1860): 21–37; in Brush, *Kinetic Theory,* 1:148–71; p. 150.
8 *Ibid.,* p. 150.
9 *Ibid.*
10 *Ibid.,* p. 170.
11 *Ibid.,* pp. 170–71.

the unqualified acceptance of the hypothesis that gases are such systems of hard elastic particles. For the ascertained fact that γ, the ratio of the specific heat at constant pressure to that at constant volume, is equal to 1.408, requires that the ratio $[\beta]$ of the whole *vis viva* to the *vis viva* of translation should be . . . 1.634; whereas, according to our hypothesis, $\beta = 2$.[12]

What then is the nature of the particles of a gas?

By the time of his next paper on the kinetic theory of gases in 1866, Maxwell had apparently become convinced of the existence of molecules and no longer uses the terms "parts" or "particles" of a gas. He now considers the molecules of a gas not as elastic spheres of definite radius but as small bodies or groups of smaller bodies repelling one another with a force inversely proportional to the fifth power of the distance. Maxwell explains that the

molecules of a gas in this theory are those portions of it which move about as a single body. These molecules may be mere points, or pure centres of force endowed with inertia, or [with] the capacity of performing work while losing velocity. They may be systems of several such centres of force, bound together by their mutual actions, and in this case the different centres may either be separated, so as to form a group of points, or they may be actually coincident so as to form one point.[13]

Thus we have yet another modification of the Boscovichean view—the point centers of force are endowed with inertia. Maxwell's explanation of the seemingly strange notion of several such points coming together to form a single point underscores his unique and profound approach to the whole problem of the nature of gaseous molecules—"The doctrines that all matter is extended, and that no two portions of matter can coincide in the same place, being deductions from our experiments with bodies sensible to us, have no application to the theory of molecules."[14]

In this paper Maxwell, for the first time, takes account of the vibrational motion of molecules. The energy of a molecule in motion is said to consist of two parts—one due to the motion of its center of gravity and the other to the motions of its parts relative to the center of gravity. If the body is rigid, the latter are rotational motions, but if the parts are not rigidly connected their motions may also consist of vibrations of various kinds.[15]

Maxwell explains that the mutual interactions of the molecules in

[12] *Ibid.*, p. 171.
[13] J. C. Maxwell, "On the dynamical theory of gases," *Phil. Trans.*, 157 (1867): 49–88; in *The Scientific Papers of James Clerk Maxwell* (Cambridge, 1890), 2:26–78; p. 33.
[14] *Ibid.*, p. 33.
[15] *Ibid.*, p. 34.

their motions among one another will cause their energy to be distributed in a certain ratio between that due to translational motion and that due to rotation and vibration. If the molecules are pure centers of force, there can be no rotation and the entire energy is that of translation; but in all other cases, the entire energy of the molecule may be represented by $\frac{1}{2}Mv^2\beta$, where β is the ratio of the total energy to the energy of translation. This ratio will be different for every molecule and will be different for the same molecule after every encounter with another molecule; but it will have an average value depending upon the nature of the molecules, as was shown by Clausius.

In the same year, 1866, Ludwig Boltzmann published the first of his many papers on the kinetic theory of gases.[16] An equation arrived at in this paper and expressing the number of atoms in a molecule as a function of the two specific heats was found not to be in agreement with experiment. For example, it was calculated that the molecules of oxygen, hydrogen, and hydrogen chloride each contain one and one-half atoms! In arriving at his equation Boltzmann had ignored the "ether masses present in a gas"; yet he did not feel that this was the source of error. Whatever this latter might be, he thought Clausius' assumption—that in elementary as in compound gases, at least two atoms are associated in a molecule—very probable.[17]

When five years later Boltzmann encountered what was essentially the same difficulty he this time attributed the source of error to the ether.[18] For diatomic molecules he calculates the ratio of the two specific heats to be 1.33, whereas for atmospheric air this ratio is found to be 1.41. The disagreement is to be explained by the fact that the gas molecules move not through empty space but through the ether, to which they lose kinetic energy by heat radiation. Nevertheless, he does allow that the true nature of gas molecules is not as yet known.[19]

It was about this time, the late 1860s and early 1870s, that physicists began to develop a qualitative molecular theory of spectra. The earliest attempt in this direction was made in 1866 by Robert Bellamy Clifton, Professor of Natural Philosophy at Owens College, Manchester. Clifton proceeds "by assuming principles closely resembling, if not identical

---

[16] L. Boltzmann, "Ueber die mechanische Bedeutung des zweiten Hauptsatzes der Wärmetheorie," *Berichte Akad. Wissen. Wien,* 53 (1866): 195–220; in *Wissenschaftliche Abhandlungen von Ludwig Boltzmann* (Leipzig, 1909), 1:9–33.
[17] *Ibid.,* p. 23.
[18] L. Boltzmann, "Über das Wärmegleichgewicht zwischen mehratomigen Gasmolekülen," *Berichte Akad. Wissen. Wien,* 63 (1871): 397–418; in *Wissenschaftliche Abhandlungen,* 1:237–58.
[19] *Ibid.,* p. 258.

with, those adopted by Professor Clausius in his well known paper on 'The Nature of the Motion which we call Heat'."[20]

"Matter is assumed in all cases to have its atoms grouped together into *molecules,* an assumption which seems necessary when the different allotropic states of certain substances are considered."[21] The molecules are assumed to be in motion, and the atoms to be vibrating with respect to one another within the molecules. "The *luminiferous ether* is supposed to exist" between the atoms of a molecule.[22]

The vibrations in the ether which constitute radiant heat and light, are considered due to the vibrations of the atoms in the molecule, and not to the motion of the molecule as a whole; the latter bearing some such relation to the ether, as a bell or a stretched string does to the air, the internal vibrations only in the two cases causing the vibrations in the surrounding media, which give rise respectively to light and sound.[23]

Due to the close proximity and consequent interaction of their molecules, continues Clifton, solids and liquids will give continuous spectra. In rarified gases and vapors, however, "the atoms will be left to vibrate under the action of the interatomic forces only, and will thus assume periods of vibration all included in a certain set; these vibrations will consequently cause vibrations in the ether corresponding only to certain definite wave lengths."[24] The resultant spectra "will be broken, and [will] consist only of a series of fine lines."

Two years later, in 1868, George Johnstone Stoney, formerly Professor of Natural Philosophy at Queen's College, Galway, expressed similar views. He describes the motions of gaseous molecules, which are held to be highly complex systems capable of being resolved into simpler entities, as being of two distinct types—the motions of molecules among one another, and the internal motions of each molecule.[25] The former account for a large part of the *vis viva;* but they are irregular, and because of this and because they are too "coarse" they are not fitted to absorb or develop vibrations in the luminiferous ether. From the mean velocity and mean free path as determined respectively by Clausius and Maxwell, Stoney calculates that on the average any gas

20 R. B. Clifton, "An attempt to refer some phenomena attending the emission of light to mechanical principles," *Proc. Literary and Philosophical Soc. Manchester,* 5 (1866): 24–28; p. 24.
21 *Ibid.,* p. 24.
22 *Ibid.,* p. 25.
23 *Ibid.*
24 *Ibid.,* p. 26.
25 G. J. Stoney, "The internal motions of gases compared with the motions of waves of light," *Phil. Mag.,* 36 (1868): 132–41.

molecule collides with another every 14×10^{-11} seconds. "This fragment of time, tiny as it is, is nevertheless more than 50,000 times the vastly shorter period that suffices for a double vibration of red light, and more than 100,000 times the duration of a double vibration of the extreme violet ray." Thus we "see the motions with which the molecules dart about amongst one another cannot produce or intercept light, in fact they are too sluggish—just as the motions of our fingers, or a very gentle waving of the hand, do not produce sound."[26]

The internal motions of a molecule, continues Stoney, are influenced by neighboring molecules only during the instants when the molecule is being deflected from its rectilinear path upon its close approach to one of the latter; and as the intervening times when the path is rectilinear are long in comparison with these brief instants, the internal motions will be for the most part undisturbed and therefore regular.

And accordingly we must presume that it is they which influence the aether, producing those bright lines which constitute nearly the whole spectrum of an incandescent gas, and are in a corresponding degree influenced by the aether, absorbing these same rays. How wonderfully regular the internal motions are, and at the same time how complex, appears to be revealed to us by the fixity of the rays in the spectrum of each gas, and by their number within the limits of the visible spectrum and the doubtless greater number that exist beyond.[27]

Stoney then explains, among other things, the broadening of the sodium D lines and the formation of a continuous hydrogen spectrum through increase of temperature, as due to the disturbing influence of neighboring molecules upon the internal vibrations.

In 1870 Maxwell himself gave a "Molecular Theory of Radiation" in his book *Theory of Heat*. Before considering this theory of radiation it is well to note that Maxwell again defines a molecule as "a small mass of matter, the parts of which do not part company during the excursions which the molecule makes when the body to which it belongs is hot."[28] But he chooses not to commit himself regarding ultimate "parts" or atoms: "We do not assert that there is an absolute limit to the divisibility of matter. What we assert is, that after we have divided a body into a certain finite number of constituent parts called molecules, then any further division of these molecules will deprive them of the properties which give rise to the phenomena observed in the substance."[29] It should be stressed that Maxwell is speaking of physical and not chemical properties.

26 *Ibid.*, pp. 139–40.
27 *Ibid.*, pp. 134–35.
28 J. C. Maxwell, *Theory of Heat* (London, 1870), p. 285.
29 *Ibid.*

Every molecule, continues Maxwell, consists of a definite quantity of matter which is exactly the same, and is bound together in the same way, for all the molecules of a given substance. A molecule may consist of "several distinct portions of matter held together by chemical bonds" and may be set in vibration, rotation, or any other kind of relative motion, and so long as the different portions remain united Maxwell regards the entire connected "mass" as a single molecule. This could well have satisfied a chemist except that nowhere in his book does Maxwell use the word "atom."

Maxwell says that "the small vibrations of a connected system [molecule] may be resolved into a number of simple vibrations, the law of each of which is similar to that of a pendulum."[30] This view will be seen in the following chapter to have important consequences. It is probable, continues Maxwell, that the gaseous molecules may execute many such vibrations in the interval between successive collisions. At each encounter the entire molecule is "roughly shaken." Then during its free path it vibrates according to its own laws, the amplitudes of the vibrations being determined by the nature of the collision, but the periods depending only upon the molecule's constitution. If the molecule is capable of communicating these vibrations to the ether then it emits certain definite radiations, and if these fall within the luminous part of the spectrum, they will be visible as light of definite refrangibility. This then is the molecular theory of spectra. Spectral lines represent the disturbances communicated to the luminiferous ether by the molecules vibrating in a regular and periodic manner during their free paths. If the free path is long, the molecule, by communicating its vibrations to the ether, will cease to vibrate until after a further collision.

Maxwell proceeds to give a qualitative explanation of known spectroscopic phenomena. Raising the temperature increases a molecule's translational velocity and hence the force of its collisions. The greater this is the greater will be the amplitude of all the internal vibrations of the molecule, and the more likely it will be that, in addition to those fundamental vibrations which are most easily produced, vibrations of short period will occur. Thus some lines will become more intense and additional lines will appear in the blue.

On the other hand, increasing the density diminishes the length of the free path of each molecule, and thus there is less time for the vibrations excited at one collision to die away before the next. Maxwell holds that since each new collision disturbs the regularity of the

<hr>

30 *Ibid.*, p. 306.

series of vibrations, the radiation will no longer be capable of complete resolution into a series of vibrations with regular periods. Rather, it will be resolved into a spectrum of bright bands produced by the regular vibrations, superimposed upon a continuous background of diffuse light produced by the irregular motion introduced at each encounter. Hence, when a gas is rarified the bright lines of its spectrum are narrow and distinct while the spaces between them are dark. As the density of the gas increases the bright spectral lines become broader and the intervening spaces more and more luminous.[31]

Throughout this Maxwell speaks as many spectroscopists of the period speak. They refer to the vibrating parts of molecules, yet never state what these parts are—whether they are atoms and if so what the nature of these is. For example, Stoney nowhere defines what he means by the "parts of a molecule." However, within a few years, Maxwell will express definite views on the nature of atoms; and as his was a dominant position, at least among British spectroscopists, it is safe to assume that they followed his lead. Later we shall see that this in fact was the case.

Maxwell's *Theory of Heat* went through a second edition in 1872, and the above views on the structure of matter and theory of spectra remained unaltered. Then in his "famous"[32] lecture on molecules given at the meeting of the British Association for the Advancement of Science in 1873, he further elaborated his ideas on matter and adopted a definite view as to the nature of the parts of molecules. The concrete nature of his views on atoms and molecules contrasts strongly with his skeptical attitude of some fourteen years earlier when he first became interested in kinetic theory.

Maxwell tells us that: "Molecule is a modern word. It does not occur in Johnson's Dictionary. The ideas it embodies are those belonging to chemistry."[33]

This needs to be qualified. "Molecule" is not a word which came into use about the middle of the nineteenth century. It had been used by, among others, Descartes, Laplace, Ampère, and Stokes, that is, for about two and one-half centuries. However, during the nineteenth century the word took on a particular meaning, a chemical meaning, as Maxwell observes. It had no longer the general meaning of "small particle" but the precise meaning which chemistry gave to it—that same meaning which today physics as well as chemistry gives to it.

---

31 *Ibid.*, pp. 306–8.
32 Lewis Campbell and William Garnett, *Life of James Clerk Maxwell* (London, 1884), p. 272.
33 J. C. Maxwell, "Molecules," *Nature*, 8 (1873): 437–41; in *Scientific Papers of J. C. Maxwell*, 2:361–67; p. 363.

Maxwell continues:

A drop of water . . . may be divided into a certain number, and no more, of portions similar to each other. Each of these the modern chemist calls a molecule of water. But it is by no means an atom, for it contains two different substances, oxygen and hydrogen, and by a certain process the molecule may be divided into two parts, one consisting of oxygen and the other of hydrogen. According to the received doctrine, in each molecule of water there are two molecules of hydrogen and one of oxygen. Whether these are or are not ultimate atoms I shall not attempt to decide.

We now see what a molecule is, as distinguished from an atom. A molecule of a substance is a small body such that if, on the one hand, a number of similar molecules were assembled together they would form a mass of that substance, while on the other hand, if any portions of this molecule were removed, it would no longer be able, along with an assemblage of other molecules similarly treated, to make up a mass of the original substance.

Every substance, simple or compound, has its own molecule. If this molecule be divided, its parts are molecules of a different substance or substances from that of which the whole is a molecule. An atom, if there is such a thing, must be a molecule of an elementary substance. Since, therefore, every molecule is not an atom, but every atom is a molecule, I shall use the word molecule as the more general term.[34]

There are several things to be remarked about this. To begin with, we see from the first paragraph that Maxwell clearly distinguishes between chemical and physical views. He does not regard the chemical atom as an atom, but as a molecule which might possibly consist of more than one "ultimate atom." Maxwell defines the latter as it had been defined for two thousand years—namely, as a body which cannot be cut in two. "One may see the atom as a material point, invested and surrounded by potential forces. Another sees no garment of force, but only the bare and utter hardness of utter impenetrability."[35] Maxwell prefers the second of these alternatives and sees himself in the company of Democritus, Epicurus, and Lucretius. But the Boscovichean view was not to be excluded.

Many chemists, on the other hand, regarded the various elemental chemical atoms as being different Lucretian-type atoms. But there were exceptions to this view and these remind us that at least a few chemists were aware that contemporary thought on atoms was of a hypothetical nature. Like Maxwell these chemists, such as Marcelin Berthelot, recognized that certain phenomena suggested rather than declared the existence of atoms and molecules. The chemists, according to Berthelot, "consider only the relations of the weights of the molecules which

34 *Ibid.*, p. 363.
35 *Ibid.*, p. 364.

combine with one another or substitute one another, their atom being defined by the minimum value of these relationships."[36] Being aware of this, as Maxwell himself was, aids us in understanding his approach. While drawing a clear distinction between physical and chemical views, Maxwell was always quick to draw attention to any common result. Thus, when he had calculated from kinetic theory considerations that equal volumes of all gases under the same conditions of temperature and pressure contain the same number of molecules, he remarked that "this result agrees with the chemical law, that equal volumes of gases are chemically equivalent."[37] We shall see during the course of the chapter that spectroscopic considerations brought the two distinct views together into a single view.

Returning now to the long quotation from Maxwell, we have no difficulty in understanding his definition of a molecule when we think of a compound molecule such as a water molecule. With a simple or elemental molecule, however, it is different. If from each molecule in a volume of an elementary substance composed of polyatomic molecules we take one or two atoms, then surely, we say, the "reduced" molecules would still "make up a mass of the original substance." But that is to think in today's, or late nineteenth-century chemical, terms. If we had recently been favorably disposed toward the Boscovichean view of matter, and in addition if we perhaps still entertained the possibility of Prout's hypothesis, then it is likely that we would regard the atoms of all substances as being identical. The physical molecules or chemical atoms of different elementary substances would then be different because they contained different numbers of the one ultimate atom. Hence Maxwell's view of elementary molecules would make sense if we assume he held all such molecules to be composed of the one type of ultimate atom. However, he does not state whether all elementary substances are composed of the same ultimate atom or whether each has a unique ultimate atom.

We at least know that the first alternative had recently been considered possible. One of the leading investigators of the physical properties of gases, Thomas Graham, stated in 1864 that it was "conceivable that the various kinds of matter, now recognized as different elementary substances, may possess one and the same ultimate or atomic molecule existing in different conditions of movement. The essential unity of matter is an hypothesis in harmony with the equal

36 M. Berthelot ["Sur l'existence réelle d'une matière monoatomique"], Comptes Rendus, 82 (1875): 1129–30.
37 J. C. Maxwell, "Illustrations of the dynamical theory of gases," in Brush, Kinetic Theory, 1:164.

action of gravity upon all bodies."[38] Graham's expression, "ultimate or atomic molecule," uses "molecule" in the old or nonchemical sense. Graham further says that the "gaseous molecule must itself be viewed as composed of a group or system of the preceding inferior atoms." This idea of a single ultimate unit of matter is met with repeatedly in the physical literature of the second half of the nineteenth century. We shall encounter it several times during the remainder of the chapter.

By way of concluding this section it is not out of place to quote some most interesting and profound remarks on the subject of chemical and physical viewpoints. These were made in 1876 by Robert Angus Smith in his preface to the volume of Thomas Graham's chemical and physical papers.

In using the word *atom* chemists seem to think that they bind themselves to a theory of indivisibility. This is a mistake. The word atom means *that which is not divided* as easily as it may mean *that which cannot be divided,* and indeed the former is the preferable meaning. Even when Lucretius speaks of primordial bodies that cannot be divided, he does not deny that they have parts, although these, as we have seen, cannot exist by themselves, and Graham, as well as other atomists, gave a similar opinion, that is, that the original atom may be far down. Graham speaks of this in more than one place. There comes to us a something indivisible by us, and it is consistent to call it an atom, as it is consistent to call the smallest particle of alum, with its 24 equivalents of water, an atom, simply because it is the smallest possible portion of alum.

Some have preferred to leave the atom without believing in infinite divisibility, and it is strange that when we come to the point of deciding on the existence of atoms the mind insists on going further and finding out of what an atom is composed. By doing this the atomist often ends by being a nonatomist, and probably this is the necessary double conclusion: We believe in atoms because Nature seems to use them, and we break them up continuously because we know not where to stop.[39]

As will be seen later, however, the nineteenth century was to ascribe "atom" to a definite level of subdivision of matter.

2. *Spectra of Chemical Compounds*

The next significant advance made in spectrum analysis had appeared as a possibility to Bunsen and Kirchhoff, yet they were unable to find a means of testing their conjecture. This was that a salt, which remained

[38] T. Graham, "Speculative ideas respecting the constitution of matter," *Phil. Mag.*, 27 (1864): 81–84; in *Chemical and Physical Researches by Thomas Graham* (Edinburgh, 1876), p. 299.
[39] *Ibid.*, p. xviii.

undissociated at the temperature of a flame, might give a spectrum different from that of the constituent metal.

At the end of their second paper on chemical analysis by means of spectral observations, Bunsen and Kirchhoff note that among the large number of salts which they have examined spectroanalytically they have not found, "in spite of the great variation in the elementary bodies combined with the metal," a single salt which did not produce the characteristic bright lines of the metal.[40] It might then be supposed, they say, that in all cases the bright lines are emitted by a metal quite independently of the other elements chemically combined with it, and that therefore the spectrum of its vapor remains the same whether the metal is free or chemically combined. Yet, they continue, this supposition is by no means founded on fact. For it is well known that the absorption lines of iodine vapor cannot be produced by hydriodic acid, and, on the other hand, that the absorption lines of nitrous acid gas are not given by a mechanical mixture of nitrogen and oxygen. Thus chemical combination does have an effect on some absorption lines, at least at low temperatures. As for higher temperatures, such as that of a "white heat," Bunsen and Kirchhoff have found nothing to indicate that there similar effects do not also occur.

From these latter considerations Bunsen and Kirchhoff conclude that two different compounds of the same metal might give two different spectra, each of which is different from that of the metal. They are aware, however, of the possibility of salts being decomposed when volatilized in the flame and thus of the lines being given in each case by the vapor of the free metal itself.[41]

A former student and subsequent colleague of Bunsen's at Heidelberg, and now Professor of Chemistry at Owens College in Manchester, Henry Enfield Roscoe was the English apostle of spectrum analysis. He lectured on the researches of Bunsen and Kirchhoff and translated several of their papers.[42] One of these was Kirchhoff's "Memoir on the Solar Spectrum and the Spectra of the Chemical Elements," in which Kirchhoff noted that in the case of the calcium spectrum bright lines which were invisible at the temperature of a coal-gas flame became

---

[40] R. Bunsen and G. Kirchhoff, "Chemische Analyse durch Spectralbeobachten," *Ann. der Physik, 110* (1860), 160–89; and 113 (1861): 337–425; trans. H. E. Roscoe, *Phil. Mag.,* 22 (1861): 329–49, 498–510; p. 510.

[41] No one seemed to think it necessary to explain why the vapor of the nonmetallic part of the compound did not also show a spectrum.

[42] For one of these translations see footnote 40. See also Roscoe's report of his address to the Royal Institution at its Friday evening meeting of 1 March, 1861—H. E. Roscoe, "On Bunsen and Kirchhoff's spectrum observations," *Notice of the Proceedings of the Royal Institution of Great Britain,* 3 (1858–62): 323–28. In this Roscoe mentions having visited Bunsen and Kirchhoff in Heidelberg during the previous summer.

visible at the temperature of an intense electric spark.[43] Roscoe and his counterpart in Natural Philosophy at Owens College, Clifton, found their attention drawn to this observation. They were able to confirm it, and also to witness similar behavior with unspecified strontium and barium salts. They further observed that not only did the new lines appear at the high temperature of an intense spark, but also that the broad bands characteristic of "the metal or metallic compound" at the low temperature of a flame or weak spark entirely disappeared. The bright lines were generally found not to be coincident with any parts of the bands. As for the "more easily reducible" alkali metals—sodium, potassium, and lithium—no deviation or disappearance of the parts of their spectra resulted from varying the temperature.

Noting the "present incomplete condition of this most interesting branch of enquiry" Roscoe and Clifton suggested a possible explanation of their observations. At the lower temperatures of a flame or weak spark the spectrum is produced by the glowing vapor of some compound, probably the oxide, of the "difficultly reducible" metal; whereas at the "enormously" high temperatures of the intense electric spark these compounds are "split up," and thus the true spectrum of the metal is obtained.[44]

Roscoe and Clifton did not pursue this subject further and their single paper appears to have generated little interest. In the same year, however, similar results were independently arrived at by Alexander Mitscherlich, who during the previous year (1861) had obtained his doctorate in chemistry at the University of Berlin, where his father was ordinary Professor of Chemistry. While engaged on some spectro-analytical work Mitscherlich chanced to notice two bright green bands associated with a substance containing barytes, which seemed to him at first to suggest the presence of a new metal. Closer investigation revealed, however, that the two green bands were given, sometimes quite alone but frequently together with the barium lines, by a solution of barium chloride in sal-ammoniac. Then with apparatus expressly designed for the purpose Mitscherlich was able to produce at will, either the barium spectrum or the two green bands. He could thereby conclude that the compound had a spectrum of its own, different from that of the metal. With strontium and calcium chlorides he also obtained spectra which were quite different from the corresponding metallic spectra.

43 G. Kirchhoff, *Researches on the Solar Spectrum and the Spectra of the Chemical Elements*, trans. H. E. Roscoe (London, pt. 1, 1862; pt. 2, 1863), pt. 1, p. 12.
44 H. E. Roscoe and R. B. Clifton, "On the effect of increased temperature upon the nature of the light emitted by the vapour of certain metals or metallic compounds," *Proceedings of the Literary and Philosophical Society of Manchester*, 2 (1860–62): 227–30; and *The Chemical News*, 5 (1862): 233–34.

These bare facts reported by Mitscherlich serve as the basis of the following, and I trust plausible, reconstruction of his process of discovery.

The importance of spectrum analysis as a tool of discovery received great emphasis with Bunsen and Kirchhoff's discovery of the two new metals caesium and rubidium.[45] This, more than anything else, is what seems to have captured the chemical imagination, and it set not a few chemists in search of further elements and personal renown. As noted above Mitscherlich was examining a "substance containg barytes" (native sulphate of barium) when he noticed the two green bands which he at first thought belonged to a new metal; and it is likely that he was in fact searching for such. Mitscherlich's next step would have been to isolate this new "metal" by chemical means. However, in attempting this he probably found that he could never obtain the two green bands except in association with the barium lines. And it is further probable that when working with a solution of barium chloride in sal-ammoniac the possibility of the two green bands belonging to undissociated barium chloride dawned upon him. Certainly someone in Mitscherlich's position at Berlin would have known of the Bunsen-Kirchhoff conjecture about chemical compounds having characteristic spectra. Mitscherlich now saw his opportunity, and being unable to obtain satisfactory results with the usual experimental arrangement, he modified it.

Spectrum analysis had hitherto suffered from the disadvantage of having no constant intense light sources. Substances held in a flame by means of a platinum wire, for example, give a transient and not very intense light. By Mitscherlich's method, however, a continuous and intense light is obtained (Fig. 1). This method employs the small piece of glass apparatus *a,* open at either end as shown. Fine platinum wires are twisted around one end of a coarser platinum wire *c,* and this end

1.

Fig. 1. Mitscherlich's Spectral Apparatus

45 R. Bunsen and G. Kirchhoff, "Découverte de deux métaux alcalins," *Bull. de La Société Chimie* (1861): 70–71.

is then jammed into the open end *b*, of the glass tube. The tube is then filled with a solution of the substance to be examined, which, by means of the capillary action caused by the fine platinum wires, is continually drawn on to the wire *c* to replace the solution vaporized in the flame. With the tubes which he employed Mitscherlich could obtain an intense light source of two hours duration.

In one tube of type *a* Mitscherlich placed a solution of one part of acetate of barytes and ten parts of ammonium acetate, and in another a solution of hydrochloric acid. In the flame the former solution gave a very intense barium spectrum, which disappeared however when the latter was also introduced into the flame. In place of the barium spectrum there appeared the two bright bands mentioned above. These same two bands were obtained more conveniently by using a mixture of one part of a concentrated barium chloride solution with twenty parts of a sal-ammoniac solution and twenty parts of hydrochloric acid containing approximately twenty per cent of hydrogen chloride. The chlorides of calcium and strontium were investigated in similar ways.

Mitscherlich next investigated the iodides of barium, calcium, and strontium, in addition to the sulphides and fluorides of barium and calcium. But obtaining no positive results he turned his attention to copper compounds, among which he found spectra characteristic of the iodide, chloride, and chlorate. With sodium and potassium chlorides he could detect no characteristic spectra, and hence concluded that neither chloride possessed such.

We recall that hitherto it was accepted that the spectrum emitted by a compound was really that of the component metal itself, but it now occurred to Mitscherlich that the spectrum might rather be that of the metallic oxide. So he investigated the oxide of sodium, and finding it to yield no spectrum concluded with surprising rashness that all spectra produced by oxygen compounds were those of the associated metals.[46]

Reports of Mitscherlich's work appeared in both English and French scientific journals, and it was by this means that early in 1863 his discoveries came to the attention of E. Diacon, a student at the Montpellier Académie des Sciences et Lettres.[47] Diacon had apparently arrived independently at one of Mitscherlich's conclusions. Believing

[46] A. Mitscherlich, "Beitrage zur Spectralanalyse," *Ann. der Physik,* 116 (1862): 499–507.

[47] Diacon gives his source as the *Répertoire de Chimie* for January, 1863. But see also "Sur les raies spectrales des combinaisons métalliques," *Bulletin de la Société de Chimie* (1862): 108–10. A translation of this latter abstract appeared in *The Chemical News,* 7 (1863): 99–100. English readers also learned of Mitscherlich's work through J. H. Gladstone's "On the violet flames of many chlorides," *Phil. Mag.,* 24 (1862): 417–19.

that for the majority of metals the absence of spectra was due to the decomposition of the chlorides in the oxidizing flame and to the non-volatile nature of the resulting oxidized products, Diacon had sought a means of volatilizing the chlorides without simultaneously causing their decomposition. In this he was successful. Then while employing his method in a spectroanalytical study of metallic chlorides Diacon noticed that calcium, and particularly barium, gave spectra very different from the accepted ones, whereas potassium gave an almost invisible spectrum and the customary blue and orange lines of strontium seemed to have disappeared. It was at this point that Diacon read of Mitscherlich's work and thus concluded with the latter that the metal chlorides have their own spectra, a conception which he noted was contrary to the accepted opinion that the electronegative element did not interfere with the radiations emitted by a volatilized salt.[48]

The essential part of Diacon's apparatus was a glass blowpipe. Chlorine gas issued from the tapered central tube, and hydrogen gas from the annular passage formed by the inner and outer tubes. The samples to be examined were placed in the interior of the flame, so as to be in the presence of an excess of chlorine and out of all contact with the surrounding air.[49]

Later in 1863 Diacon reported that, and his own words best express his particular way of viewing the matter, "bromine, iodine and fluorine also change the nature of the rays emitted by a metal when they form with it a stable volatile compound."[50]

In 1864 this extended view of the origins of spectra received further confirmation from Hendrick Cornelius Dibbits, an assistant in the chemical laboratory at Utrecht, and also from Mitscherlich, who was now a *Privatdozent* in chemistry at the University of Berlin.[51] Mitscherlich's work is most impressive. Following his previous communication he investigated the spectra of all the metals available, together with many of their compounds, and from the results obtained he concluded "that every compound of the first order which is not decomposed, and which is heated to a temperature adequate for the produc-

48 E. Diacon, "De l'emploi du chalumeau à chlorhydrogène pour l'étude des spectres," *Comptes Rendus*, 56 (1863): 653–55. Also in the "Extraits des procès-verbaux de l'Académie des Sciences et Lettres de Montpellier" for April, 1863, *Mémoires de la Section des Sciences de l'Académie des Sciences et Lettres de Montpellier*, 5 (1861–63): 441–42. (Referred to below as *Mémoires . . . de Montpellier*.)

49 E. Diacon, "Recherches sur l'influence des éléments électro-négatifs sur les spectres des métaux," *Mémoires . . . de Montpellier*, 6 (1864–66): 129–39.

50 E. Diacon, "Recherches spectrometriques," in "Extraits des procès-verbaux . . .," *Mémoires . . . de Montpellier*, 5 (1861–63): 449–50.

51 H. C. Dibbits, "Ueber die spectren der flammen einiger gase," *Ann. der Physik*, 122 (1864): 497–545.

tion of light, exhibits a spectrum peculiar to this compound, and independent of other circumstances."[52]

Mitscherlich now recognized, contrary to his previous declaration, that several oxides do exhibit characteristic spectra. As examples of metals whose compounds become decomposed before being adequately heated, and hence give only the spectrum of the metals themselves, he mentions the alkali metals, magnesium, zinc, cadmium, silver, and mercury. He also points out that a comparison of the various spectra shows that the metallic spectra consist of individual sharp lines, whereas the spectra of their compounds with the metalloids (excepting the haloid salts of calcium, strontium, and barium, whose spectra consist of individual lines) are composed of "broad luminosities with narrow dark lines which recur at definite intervals."[53] Thus, while the terminology was not yet in use, we have here the first clear distinction between line and band spectra.

In his experiments Mitscherlich employed eight different methods, six employing flames and two the electric spark. Among the six were the method of his first paper and one similar to that used by Diacon.

Diacon published a further paper in 1865 in which he restated his agreement with Mitscherlich, while adding little of novelty to the subject.[54]

3. Multiple Spectra: Real or Unreal?

The initial step in the next major advance in spectrum analysis was also made in Germany. This was the quite unexpected and seemingly contradictory discovery by Julius Plücker and Johan Wilhelm Hittorf that some elements have not one, but two, characteristic spectra. In sharp contrast to the ready acceptance of compound spectra, the reality of double or—as it was soon to be—multiple spectra was debated over several years. The case for multiple spectra was advocated on rational experimental grounds and largely against a less rational and theoretical opposition which found strength in the still unsophisticated state of spectral and molecular knowledge. The opposition held tenaciously to the one-substance-one-spectrum principle which was the very basis of spectrum analysis, and they were prepared to ascribe any additional spectra to the presence of impurities. When the issue had finally been

---

[52] A. Mitscherlich, "Ueber die spectren der verbindungen und der einfachen körper," *Ann. der Physik,* 121 (1864): 459–88; *Phil. Mag.,* 28 (1864): 169–89; p. 176.
[53] *Ibid.,* p. 182.
[54] E. Diacon, "Recherches sur l'influence des éléments électronégatifs sur le spectre des métaux," *Annales de Chimie et de Physique,* 6 (1865): 5–25.

decided, spectrum analysis instead of being undermined was more sophisticated—as was also the physicist's knowledge of matter.

Plücker had been one of the first to study the spectra of gases. As early as 1858 he had recognized that each gas has a "characteristic spectrum."[55] In the following year he observed that the spectrum of a mixture of gases is an overlapping of the spectra of the individual gases in the mixture.[56] By 1859 he had also recognized that in Plücker tubes "gases which are composed of two simple gases (aqueous vapour, ammonia, nitrous oxide, nitrous acid) immediately split up into their simple constituents" thus giving a spectrum consisting of two overlapping spectra.[57] But most important, Plücker noted that gaseous spectra "are essentially different from those belonging to the electrical arch of light, and from metals glowing or burning in it";[58] and further, that the spectra of chlorine, bromine, and iodine are different from all other gaseous spectra in that they consist of *"lines of light"* whose width is about the same as that of the narrow Fraunhofer lines.[59] In gaseous spectra "the colours do not merge into one another as in the ordinary solar spectrum. They are, on the contrary, sharply demarcated; and the separate spaces of colour again are also divided into well defined lighter and darker stripes."[60]

Thus gaseous spectra are composed of bands rather than bright lines, a fact which is readily illustrated by the descriptions of two gaseous spectra given by Plücker in 1858. One of these is the hydrogen spectrum which consists of a bright violet band, three bands in the green, and a "beautiful" yellow band.[61] The second is the spectrum of the vapor of the compound boron fluoride. It has bright bands in the red, and in the blue has the appearance of a fluted column.[62]

55 J. Plücker, "Ueber die Einwirkung des Magneten auf die electrische Entladung in verdünnten Gasen," *Ann. der Physik*, 104 (1858): 113–28; in *Phil. Mag.*, 16 (1858): 408–18; p. 414.

56 *Ibid.*, 105 (1859): 67–84; in *Phil. Mag.*, 18 (1859): 7–20; p. 15.

57 *Ibid.*, p. 19. In what follows Geissler as well as Plücker tubes will be referred to often. Heinrich Geissler was a manufacturer of scientific instruments in Bonn, Germany. In 1855 he invented a vacuum pump which made use of the Torricellian vacuum. The latter was used to draw gas from the vessel to be evacuated and then re-created. Continuous repetition of this cycle eventually produced a vacuum in the vessel comparable to the Torricellian vacuum. Tubes evacuated in this way were called Geissler tubes by Plücker, an associate of Geissler's. Plücker himself designed a tube similar in shape to an hour-glass but with a relatively long capillary section in the middle. This became known as a Plücker tube. See H. Schellen, *Spectrum Analysis*, trans. Jane and Caroline Lassell (New York, 1872), pp. 24–27.

58 J. Plücker, "Ueber die Einwirkung des Magneten auf die electrische Entladung in verdünnten Gasen," p. 418.

59 *Ibid.*, *Ann. der Physik*, 105 (1859): 67–84; in *Phil. Mag.*, 18 (1859): 7–20; p. 15.

60 *Ibid.*, *Phil. Mag.*, 16 (1858): 414.

61 *Ibid.*, p. 415.

62 *Ibid.*, pp. 416–17.

That Plücker continued these researches during the year 1860 is uncertain, but in the summer of the following year he was joined by Hittorf, then Professor of Physics and Chemistry at the Academy of Münster, and together they investigated gaseous spectra. One of their objectives was to observe the behavior of spectra at high temperatures, produced by passing heavy currents through Plücker tubes. To their surprise they discovered that in addition to the familiar increase of intensity with increase of temperature some spectra actually became transformed. The beautiful band spectrum of nitrogen, for example, which had first been obtained by Plücker in using a small Ruhmkorff apparatus, remained essentially unaltered when a large Ruhmkorff apparatus was substituted. But when the latter was used, together with a Leyden jar, an entirely new spectrum showing no traces of the original band spectrum was given. On the contrary, it was composed of a great number of beautiful lines. The vapors of sulphur and selenium were found to behave in a similar manner.[63]

Plücker's immediate concern was to elucidate the meaning of this discovery for spectrum analysis,[64] and it was only several months later —in January 1863—that he formally announced the idea of double spectra.[65] As it contradicted the fundamental principle of spectrum analysis, the delay in making this "remarkable and at first sight improbable assertion"[66] is quite understandable. In the words of another spectroscopist the Plücker–Hittorf discovery was the most startling addition to spectrum analysis in the ten years following its inception.[67] A measure of its import can be had by considering William Crookes's remark in connection with the discovery of thallium two years earlier, namely, that if an element can be so changed as "to have the spectrum of its incandescent vapour (which is, *par excellence,* an elementary property) altered to an appearance totally unlike that given by its former self, it must have been changed into something which it originally was not."[68]

The only full account of the Plücker–Hittorf work was submitted, curiously, to the Royal Society of London in February 1864. The nitrogen spectra still afforded the best illustration of the phenomenon.

Nitrogen in the state of greatest rarefaction, which as may be obtained in Geissler's exhauster, like other gases does not allow the induction current to pass through. But when its tension is only a small fraction of a millimetre, the

---

63 J. Plücker, "Analyse spectrale," *Cosmos,* 21 (1862): 283–88.

64 *Ibid.,* pp. 312–15.

65 Reported in *Decheniana,* 20 (1863): 38–42.

66 A. Schuster, "Spectra of metalloids," *Rep. British Assn.,* 49 (1880): 258–74; p. 258.

67 Wm. Marshall Watts, "On double spectra," *Quarterly J. Sc.,* 1 (1871): 1–15; p. 6.

68 Quoted in E. E. Fournier D'Albe, *The Life of Sir William Crookes O.M., F.R.S.* (New York, 1924), p. 62.

current begins to pass and renders the gas luminous. Below a certain limit of temperature ignited nitrogen sends out a golden-coloured light, giving the spectra of bands. Above this limit the colour of the light is replaced by a bluish violet, the spectrum of chanelled spaces replacing simultaneously the spectrum of bands. When, by means of the intercalated jar for instance, the temperature rises to a second higher limit, the light of the gas, becoming white and almost brilliant, gives, if analyzed by the prism, a spectrum of quite a different description [elsewhere they indicate that the transition is a discontinuous one]: bright lines of different intensity, with the colour indicated by the place they occupy, rise from a dark ground. By increasing the power of the discharge these lines become more brilliant, but the brilliancy does not increase in the same ratio for them all. New bright lines appear, which formerly, on account of their extreme faintness, were not visible, but the number of such lines is not unlimited. By increasing the heat of the ignited nitrogen to the last extremity, the lines, especially the brighter ones, gradually expand, approaching thus to a continuous spectrum.[69]

Thus nitrogen successively exhibits, as its temperature increases, two band spectra and one line spectrum. Sulphur and selenium are said to have two spectra—a line and a band spectrum; but oxygen is found to have only one.[70] Hydrogen displays two spectra: one showing the three bright lines denoted by Plücker as H_α, H_β, and H_γ; and the other "corresponding to a lower temperature, but having no resemblance at all" to the band spectra of hydrogen, sulphur, and selenium. This latter "peculiar" spectrum displays numerous well-defined bright lines in the red and yellow.

Plücker and Hittorf also examined the transformation of compound spectra into elemental spectra, a change which they also described as the replacement of a band by a line spectrum. This close similarity of compound and double spectra is also to be seen in the work of Mitscherlich, who would seem to have discovered the latter independently in 1864. Before considering this, however, I would suggest that Mitscherlich's first paper on compound spectra could have led Plücker and Hittorf to the discovery of double spectra.

Mitscherlich's paper was published in Poggendorf's *Annalen* at the end of July or beginning of August, 1862, while Plücker's first communications were in the form of letters published in the journal *Cosmos* on September 12 and 19 of the same year. Mitscherlich had described the effect of increased temperature upon the spectra of the incandescent vapors of compounds, and this about the time when Plücker and Hittorf began to investigate the effect of increased temperature upon gase-

69 J. Plücker and W. Hittorf, "On the spectra of ignited gases and vapours, with especial regard to the different spectra of the same elementary gaseous substance," *Phil. Trans.*, 155 (1865): 1–29; p. 13.
70 *Ibid.*, pp. 13–15, 23.

ous spectra. Most important, Plücker had known for several years that vapors of compounds give spectra of bands. It is clear what course Mitscherlich's work would have dictated. And if the band spectra of compound vapors could be transformed, what then of the band spectra of elementary gases?

For his own part Mitscherlich noted in 1864 that the metalloids give the same type of shaded band spectra as the metallic oxides.[71] At high temperatures, however, iodine exhibits an entirely different spectrum consisting of individual bright lines similar to those produced at high temperatures by metallic oxides. Bromine likewise has two different spectra, one an absorption spectrum and the other produced in the electric discharge. Then prompted by these examples and finding similarities to the flame spectrum of iodine in the flame spectra of selenium, tellurium, and phosphorus, and in the discharge spectra of sulphur and nitrogen, Mitscherlich infers that any metalloid would behave as iodine. Finally, in a note of the following year he states that experiment has shown almost all metalloids to possess double spectra and promises an account of the experiments in the near future.[72]

Nothing more was heard from Mitscherlich, however, on the subject of multiple spectra, nor for that matter from Plücker and Hittorf. Plücker died in 1868. As for Hittorf his interest in gas-discharge tubes led him to the discovery of cathode rays in 1869 and away from gaseous spectra. Even the founders of spectrum analysis, Bunsen and Kirchhoff, remained silent on this new and contradictory concept which Plücker and Hittorf had considered "important, as well as with regards to theoretical conceptions as to practical applications."[73] In fact double spectra were not discussed in print again until 1868, when a lively debate was begun.

It was another German physicist, Adolph Wüllner, who in 1868 revived interest in multiple spectra and initiated the debate on their existence. Curiously, Wüllner was Plücker's successor in physics at the University of Bonn, having been appointed in the previous year. During an experimental investigation of the relationship between refractive index and density, in which he was using Plücker's hydrogen spectrum of three bright lines, Wüllner accidentally observed a third hydrogen spectrum. We recall that Plücker's second spectrum was a band spec-

---

[71] A. Mitscherlich, "Ueber die spectren der verbindungen und der einfachen körper," p. 176.
[72] A. Mitscherlich, "Ueber die Anwendung der Verbindungenspectren zur Entdeckung von Chlor, Brom und Iod in geringsten Menge," *Ann. der Physik,* 125 (1865): 629–34; p. 634.
[73] J. Plücker and W. Hittorf, "On the spectra of ignited gases and vapours," p. 6.

trum which, while corresponding to a lower temperature than the line spectrum, showed no similarity to any other band spectrum. Wüllner's third hydrogen spectrum, however, had decidedly the character of a band spectrum, especially in the green.[74] But might not this spectrum be due to some impurity? Wüllner obtained some hydrogen tubes from Geissler. And one of these readily gave the third spectrum "so rich and beautiful in its shading." Furthermore, following the lead of Plücker and Hittorf and employing a Holtz machine and Leyden jar Wüllner could obtain the line spectrum of three bright lines with the same tube.

Since the hydrogen spectra appeared at different temperatures, which were taken to depend on the mode of discharge employed, Wüllner argued that the different spectra should also be given when one mode of discharge, or one temperature, was maintained and the pressure of the gas varied. This he was able to confirm in collaboration with Anton Bettendorff, a citizen of Bonn who earlier had been the first assistant in the chemical laboratory at the town's university. As the pressure of the hydrogen gas was continuously reduced by means of a Sprengel pump they passed successively from a continuous spectrum to Wüllner's band spectrum to the spectrum of three bright lines.[75]

When the gas was further rarified to the limit of the pump yet another spectrum was obtained. This consisted principally of six "splendid" groups of green lines. The presence of this spectrum was "so surprising" that Wüllner thought it might be due to an impurity in the gas. Possible impurities were: aluminum from the electrodes, mercury from the vacuum-pump, phosphorus and sulphur from the phosphoric and sulphuric acids used to dry the gas, and carbon from evaporated stopcock grease. However, when the spectra of these several elements were compared with the new hydrogen line spectrum the latter was seen to be different from each of them.[76] Thus Wüllner now recognized four different hydrogen spectra, none of which was the band spectrum recorded by Plücker and Hittorf. These were: a continuous, a band, and two line spectra.

Oxygen was found in the same way to have three distinct spectra, depending upon the pressure of the gas.[77] With nitrogen, however, Wüllner found no more than the two spectra noted by Plücker and Hittorf.[78]

---

[74] A. Wüllner, "Ueber die Spectra einiger Gase in Geissler'schen Röhren," *Ann. der Physik*, 135 (1868): 497–527; in *Phil. Mag.*, 37 (1869): 405–24; p. 407.
[75] *Ibid.*, p. 411.
[76] *Ibid.*, p. 416.
[77] *Ibid.*, pp. 417–22.
[78] *Ibid.*, pp. 423–24.

Whereas Plücker and Hittorf's announcement of double spectra had drawn no public response, it was otherwise with Wüllner's work. In the same year in which it appeared Ångstrom and Augustin Pierre Dubrunfaut, Professor of Industrial Chemistry at the trade school in Paris, each rejected the existence of multiple spectra.

In 1868 Ångstrom published his *Recherches sur le Spectre Solaire* which has been called his greatest work and which is said to have met with recognition and admiration from all sides.[79] To this he appended a text in which he states that he and his assistant, Tobias Robert Thalén, have not been able to confirm Plücker's discovery of double spectra with an "elementary body." On the contrary, as the temperature is raised it is found that the intensities of the lines vary in a very complicated manner, and that even new lines may appear if the temperature is sufficiently raised. But independently of these changes, the spectrum of a given body will always maintain its individual character. Therefore "any" fluted spectrum appearing with an elementary body is due to an impurity.[80]

For his part Dubrunfaut asserts that in examining simple gases such as oxygen, hydrogen, and nitrogen spectroanalytically, he has come to recognize that no matter how carefully they are prepared they always contain impurities. Thus, for example, traces of nitrogen are always to be found in the purest samples of hydrogen and oxygen.[81] Dubrunfaut then argues that these findings could explain the observations which have led several savants, and he mentions Plücker and Wüllner, to admit multiple spectra for these two gases.

Dubrunfaut does not find it so easy to establish the nature of the impurity in nitrogen prepared by heating air over copper. However, when observed under various conditions of temperature and pressure he finds that this gas gives only the band spectrum. This makes an interesting contrast with Ångstrom's conclusion.

In dealing with the hydrogen spectra Dubrunfaut attributes not only the two spectra observed by Plücker but also the two observed by Wüllner, to impurities, thus for the moment allowing hydrogen to have no spectra at all! This example serves to show a lack of care often in evidence in Dubrunfaut's work. Finally, he vaguely ascribes the "so-called" multiple spectra of the three gases to alterations in the electrodes and to mercury vapor diffusing from the Sprengel pump.

79 See Anna Beckmann, "Anders Jonas Ångstrom," in *Swedish Men of Science,* ed. Sten Lindroth (Stockholm, 1952), pp. 193–203; p. 202.
80 A. J. Ångstrom, *Recherches sur le Spectre Solaire* (Uppsala, 1868), pp. 38–39.
81 A. P. Dubrunfaut, "Essai d'analyse spectrale appliquée a l'examen des gaz simples et de leurs mélanges," *Comptes Rendus,* 69 (1869): 1245–49; p. 1246.

Thus according to Dubrunfaut an elementary gas has only one characteristic spectrum.[82]

Dubrunfaut used spectral tubes constructed after Plücker's design and which he himself had filled. That he could never prepare hydrogen or oxygen tubes free of nitrogen indirectly says much for Geissler, whose tubes were used by both Plücker and Wüllner. Dubrunfaut found, for example, that a one-thousandth part by volume of nitrogen in hydrogen was sufficient to produce a complete nitrogen spectrum just as brilliant as the hydrogen spectrum.[83]

In reply to Dubrunfaut, Wüllner states that he seems only to have read an abstract of his work, and that if he had known the details of Wüllner's observations Dubrunfaut would probably have arrived at different conclusions.[84] Wüllner also observes that as Dubrunfaut ascribes the band spectra of both hydrogen and oxygen to nitrogen he must therefore admit that nitrogen when mixed with hydrogen gives a different spectrum than when mixed with oxygen, and yet another spectrum when pure. For these three spectra are observed to be different from one another. Furthermore, Wüllner takes Dubrunfaut to have said that the new line spectra of hydrogen and oxygen are in reality due to mercury, and replies that Dubrunfaut could easily test this "hypothesis" for himself by comparing the wavelengths given by Plücker for mercury with those given by Wüllner for hydrogen and oxygen.

Dubrunfaut's opinion, however, remained unchanged, and within two months he again disputed the existence of multiple spectra. This time Wüllner did not bother to reply. Nevertheless, it is of interest to briefly indicate the substance of Dubrunfaut's paper.

Dubrunfaut mentions a recent observation made by Pietro Angelo Secchi, the Jesuit director of the Gregorian University Observatory in Rome and an energetic astrophysicist. While experimenting with a Plücker tube containing rarified nitrogen, Secchi observed three different spectra appearing simultaneously. These were: the spectrum of brilliant lines in the capillary part of the tube; and, in the wider parts, two channelled spectra—one similar to that observed by Plücker and the other of much broader flutings.[85] Dubrunfaut states that he had independently made similar observations. For example, he was "not a little" surprised to notice that the luminous point of the positive

82 *Ibid.*, p. 1248.
83 *Ibid.*, p. 1246.
84 A. Wüllner, "Sur les spectres des gaz simples," *Comptes Rendus,* 70 (1870): 125–29.
85 P. A. Secchi, "Sur la constitution de l'auréole solaire, et sur quelques particularités offertes par les gaz raréfiés, lorsqu'ils sont rendus incandescent par les courants électriques," *Comptes Rendus,* 70 (1870): 79–84.

electrode in a Plücker tube containing hydrogen displayed only a band spectrum without the characteristic lines of hydrogen. The blue aureole of the negative electrode, on the other hand, displayed at the same time a distinct hydrogen spectrum. Moreover, similar appearances were seen with oxygen and nitrogen tubes.[86]

These "surprising" results are once more explained, however, in terms of impurities. In the case of hydrogen the positive electrode, being always at a lower temperature than the negative electrode, produces that spectrum which according to Dubrunfaut first appears at lower temperatures, namely the spectrum of the impurity, nitrogen. The negative (he actually writes "positive") electrode, on the contrary, because of its higher temperature, produces the hydrogen spectrum.

It is sufficient merely to add Dubrunfaut's concluding remarks, as they epitomize his general approach and attitude:

After having thus refuted the existence of the second-order [line] hydrogen spectrum, ought we to insist on the probability of the non-existence of the other spectra? We have pointed out in the lack of purity a cause of errors which is common to all simple gases and independent of the pressure, and it is evident that this permanent cause of errors must affect all the results of experiments which have been made without taking them into account.[87]

It is easily understood why Wüllner attempted no reply. To him this was no way to carry on scientific research. He had already made clear that the only legitimate means of deciding upon the existence or nonexistence of multiple spectra was to make a comparison of spectra. This he had done himself and was convinced that multiple spectra were a reality.

However, he was soon led by this very means to reject two of his previously declared multiple spectra as spurious. Having recognized the band spectrum, which he had ascribed to oxygen, as being most similar to one of the spectra of carbon described by William Marshall Watts, and also to the spectrum of carbon monoxide obtained by Plücker in highly rarified Geissler tubes, Wüllner made a thorough study at various pressures of different gaseous hydrogen, oxygen, and nitrogen compounds of carbon.[88] He confirmed the authenticity of the hydrogen spectra, but the band spectrum and the new line spectrum of oxygen were found to be due to the presence of carbon compounds.[89]

86 A. P. Dubrunfaut, "Sur les spectres de divers ordres des corps simples," *Comptes Rendus*, 70 (1870): 448–51.
87 *Ibid.*, p. 450.
88 W. M. Watts, "On the spectra of carbon," *Phil. Mag.*, 38 (1869): 249–63; and J. Plücker and W. Hittorf, "On the spectra of ignited gases and vapours," pp. 18–19.
89 A. Wüllner, "Ueber die Spectra einiger Gase in Geissler'schen Röhren," *Ann. der Physik*, 144 (1872): 481–525.

Wüllner's method of examining suspected multiple spectra had meanwhile been adopted by Ångstrom who, however, maintained his former position in arriving in 1871 at the opposite conclusion to Wüllner's.[90] Ångstrom's paper was published just a few weeks before Wüllner withdrew the two spurious oxygen spectra.

Ångstrom remarks that the question of multiple spectra is a vital one for spectrum analysis and that in this light Wüllner's observations are truly important. Before commencing to analyze the phenomena he cautions that the results obtained with gases are not absolutely certain when the rarefaction is carried to its utmost limits. With a tube containing atmospheric air in such a state he once successively obtained the following spectra: the ordinary air-spectrum; the fluted spectrum of nitrogen; the spectrum of carbonic oxide; and, when the rarefaction was maximum, the lines of sodium and chlorine. Furthermore, when a mercury pump is employed, the lines of mercury may appear, just as those of sulphur may appear when sulphuric acid is used to dry the gas. Thus the outcome may easily be a multiplicity of spectra which it would be incorrect to attribute to one and the same gas.[91]

Then for hydrogen, Ångstrom in turn establishes Wüllner's continuous spectrum as the ordinary spectrum of hydrogen; his band spectrum as that of acetylene; and the new line spectrum as that of sulphur. This last fact Ångstrom believes to be "most positively demonstrated" from a comparison with the wavelengths of the sulphur lines. Thus Ångstrom maintains that hydrogen has but one spectrum.

As regards oxygen, Ångstrom has reproduced and studied Wüllner's "supposed" second oxygen spectrum, that is, the band spectrum, and has found it very similar to the carbon monoxide spectrum. Ångstrom also demonstrates that the new line spectrum of oxygen is really that of chlorine.

Regarding the spectrum attributed to nitrogen, Ångstrom only mentions in a footnote his conviction that the fluted bands so characteristic of metallic oxides are never found among the spectra of elementary gases.[92] He reiterates his belief in the "absolute" principle that an elementary gas has a single characteristic spectrum of bright lines. The number of lines may increase in number and their relative brightness vary with increasing temperature, but the spectrum nevertheless maintains its character unaltered.

In reply Wüllner points out that, assuming the rate of emission of

90 A. J. Ångstrom, "Sur les spectres des gaz simples," *Comptes Rendus*, 73 (1871): 369–73; in *Phil. Mag.*, 42 (1871): 395–99.
91 *Ibid.* (translation), pp. 395–96.
92 *Ibid.* (original), p. 370. (Not included in translation.)

light to be constant for all temperatures and pressures, he has shown that the spectra of gases are continuous at high temperatures, a result which follows directly from Kirchhoff's law.[93] Likewise Friedrich Zöllner has demonstrated that for an assumed constant rate of emission the distribution of light in a spectrum can change completely with increasing thickness of gas.[94] And yet, says Wüllner in gently mocking terms, Ångstrom believes that a spectrum cannot change its character.

It should be said here that there were two ideas which militated against the acceptance of multiple spectra by the physicist. First, and perhaps more important, the fundamental principle of spectrum analysis taught that each elementary body had only one spectrum. For such a body to have more than one spectrum it would necessarily have to have more than one molecular form. However, and this is the second idea, the physicist regarded all molecules of the one element as being identical. The precise notions of atom and molecule which were to crystallize later had hardly begun to take form, as was seen in section 1 of this chapter. Rather, the nebulous term "molecule," meaning the smallest particle of matter encountered in physical phenomena and yet not necessarily an ultimate atom, had sufficed for the physicist's needs.

The chemists, on the other hand, had traditionally, although not exclusively, always spoken in terms of atoms. However, under the combined influence of Cannizzaro and Avogadro they had recently introduced the very useful hypotheses of monatomic, diatomic, etc., elementary molecules. Thus molecules were very definite entities in the chemistry of the early 1870s. Chemical hypotheses, however, were conceived with the object of explaining chemical phenomena. They did not necessarily have any application to physical phenomena.

Some elementary bodies, however, were known to have allotropic forms. And while allotropy was as yet little understood, a physicist could suggest, as Plücker and Hittorf had suggested with nitrogen, that the different spectra of an elementary body were due to different allotropes.[95] Yet while some few elements were known to have allotropes the great majority were regarded as having none. Certainly hydrogen and nitrogen, for example, for which multiple spectra were being claimed, had no known allotropic forms.

For many physicists these factors must have greatly reduced the possibility of multiple spectra occurring. And in addition, as Ångstrom

[93] A. Wüllner, "Ueber die Spectra einiger Gase in Geissler'schen Röhren." Wüllner's reply is on pp. 520–25.
[94] F. Zöllner, "Ueber den Einfluss der Dichtigkeit und Temperatur auf die Spectra glühender Gase," *Ann. der Physik*, 142 (1871): 88–111.
[95] J. Plücker and W. Hittorf, "On the spectra of ignited gases and vapours," p. 10.

and others had pointed out, there was the ever-present problem of contamination by impurities.

Returning now to Wüllner's reply to Ångstrom, Wüllner rejects the latter's criticism that the high pressure continuous hydrogen and oxygen spectra are respectively identical to the hydrogen and oxygen line spectra. Wüllner next considers Ångstrom's contention that the new line spectrum of hydrogen is really the spectrum of sulphur. In the line spectrum of sulphur described by Plücker, he says, there is such a multitude of lines in the green and blue that an approximate coincidence with the green lines of another spectrum cannot be significant. Wüllner specifies an "approximate" coincidence, for the differences in the positions of the lines of his hydrogen spectrum and those of sulphur amount significantly in some instances to several minutes. In illustrating how little such an approximate conformity of lines proves when only eight of a group of lines are considered, Wüllner at the same time counters Ångstrom on a second point. On such evidence Ångstrom has ascribed Wüllner's second line spectrum of oxygen to chlorine, and while Wüllner has shown that this spectrum really belongs to carbon, it nevertheless agrees better with the chlorine spectrum than the hydrogen spectrum agrees with that of sulphur. Wüllner continues that if one wishes to identify a spectrum from a coincidence of lines, then the coincidence must be a complete and not a partial one. On this basis he implores Ångstrom to again compare the new line spectrum of hydrogen with that of sulphur.[96]

Meanwhile, in the same year, 1872, the behavior of nitrogen was closely investigated by Arthur Schuster, then a student under Roscoe and Balfour Stewart at the Physical Laboratory of Owens College, Manchester.[97] Schuster, and presumably also Roscoe and Stewart, preferred Ångstrom's conclusions to Wüllner's on the question of multiple spectra. With the respected Ångstrom replying to Wüllner on his own terms, and being supported by others, Wüllner's solitary position seemed weakened.

The objectives of Schuster's research were to establish that pure nitrogen gives only the line spectrum and that the fluted spectrum is due to oxides of nitrogen produced in the electric discharge.

In repeating Secchi's experiment on nitrogen Schuster initially could find only the line spectrum as opposed to the three spectra observed simultaneously by Secchi. Yet after a time he obtained the band spectrum, which could be transformed into the line spectrum by employing an appropriate discharge. After further time had elapsed, the dis-

---

96 A. Wüllner, "Ueber die Spectra einiger Gase in Geissler'schen Röhren," p. 524.
97 A. Schuster, "On the spectrum of nitrogen," *Proc. Roy. Soc.*, 20 (1872): 484–87; and *Phil. Mag.*, 44 (1872): 537–41.

charge ceased to pass as the tube had been leaking. Schuster therefore concluded that the presence of air was necessary for the formation of the band spectrum. It was apparently well known that oxides of nitrogen are formed by the electric discharge passing through air. And the similarity of the nitrogen band spectrum and the spectra of metallic oxides seemed to Schuster to render his view probable.

Schuster consequently devised experiments by means of which he demonstrated, first, that the fluted nitrogen spectrum appeared when traces of oxygen were present, and second, that in the absence of oxygen the nitrogen line spectrum appeared under all conditions of temperature and pressure. He employed a modified Plücker tube filled with nitrogen in which small pieces of sodium were placed. The discharge was first passed through the tube without the sodium having been heated, whereupon the fluted spectrum appeared. The sodium was next heated until it presented a clean metallic surface, at which point the nitrogen was considered to be entirely free of oxygen. When the discharge was now passed, the line spectrum with its characteristic green line appeared. Schuster therefore concluded that the fluted bands belonged not to pure nitrogen but to one of its oxides. Which one he was unable to determine, but he thought it was probably nitric oxide.

In the same year Schuster also published a note on the spectrum of hydrogen.[98] His experiments, he says, confirm Ångstrom's opinion that the so-called band spectrum attributed to hydrogen by Plücker is really that of acetylene. Hydrocarbons, he explains, may be introduced into the tube by the hydrogen itself carrying with it small pieces of rubber on its passage through the india-rubber tubing. Also, the spectral tubes are "more or less greasy." These two causes are sufficient in Schuster's view to produce all of the effects observed by Plücker. In Wüllner's case the source of error is to be found in the grease on the stopcocks of the spectral tube.

Not having a Sprengel pump, Schuster could not attempt to obtain Wüllner's second hydrogen line spectrum. He observes, however, that it had not been obtained by Plücker and also that Ångstrom has attributed it to sulphur. Then mentioning Wüllner's reply to Ångstrom —that the sulphur spectrum is different from this hydrogen spectrum— Schuster counters with: "But this may be due to the circumstance that the sulphur-spectrum was never examined under so minute a pressure [as this hydrogen spectrum]."[99] His loyalty is undivided—or so it would seem.

[98] A. Schuster, "On the spectrum of hydrogen," Nature, 6 (1872): 358–60; and Rep. British Assn. 41 (1872): 38–39 (Section).
[99] A. Schuster, "On the spectrum of hydrogen," Nature, p. 359.

But in his concluding remarks Schuster draws a most important distinction between two "different subjects" which he considers to be confused with one another.

> We have first bodies which are gases at the ordinary temperature, such as nitrogen, hydrogen, oxygen. The question, whether these bodies can give different spectra under different circumstances, must, I think, be answered in the negative. This was the opinion expressed by Ångstrom from the beginning; and although this physicist clearly obtained all the results mentioned by Plücker and Wüllner, they seem always to have left the conviction in his mind, that they are due to impurities. But there are other bodies, such as iodine, sulphur and bromine. The existence of two spectra, in the case of iodine and sulphur, seems to be satisfactorily established by Mr. Salet. One of these spectra is the reversal of the absorption bands of the vapour of these bodies. While the others are spectra of lines.[100]

Thus double spectra were given by some elements, but not by ordinary gases. It is significant that all of the former were regarded as having allotropic states, as we now see from Salet's work.

Georges Salet was a chemist and assistant to Adolphe Wurtz, Professor of Chemistry at the École de Médecine and the Sorbonne in Paris. Salet published the results of his spectral investigations of sulphur in 1871.[101] Mentioning the various potential sources of impurity associated with Geissler tubes, he remarks that it can be readily understood why Ångstrom rejected the three additional spectra attributed to hydrogen by Wüllner. Nevertheless, Salet finds that Plücker's discovery is "left unshaken," at least as regards sulphur, which was known to have allotropic states. For he has found two perfectly distinct and characteristic sulphur spectra—one of lines, the other of bands. The first is produced by the disruptive discharge, whereas the second may be produced either as an absorption spectrum, or as an emission spectrum when a discharge of lower electrical potential or the hydrogen flame is employed.

Another important feature of Salet's work is his modified Plücker tube designed to eliminate any spurious effects caused by electrodes or stopcock grease. The electrodes did not pass through the walls of the tube into its interior, but instead took the form of external metallic sheaths around either end of the tube. There were no stopcocks. The sheaths were connected in the usual way to an induction coil or Holtz machine and by influence the tubes became just as luminous as tubes with internal electrodes.

Salet soon reported that selenium, whose properties are similar to

[100] *Ibid.*, pp. 359–60.
[101] G. Salet, "Sur le spectre d'absorption de la vapeur de soufre," *Comptes Rendus,* 73 (1871): 559–61. See also, Salet, "Sur le spectre d'absorption de la vapeur de soufre," *Comptes Rendus,* 74 (1872): 865–66.

those of sulphur and which also has allotropic forms, likewise displays two emission spectra.[102] Again a sheathed tube was used.

Then when experimenting with iodine in the following year (1872), Salet found in addition to a line spectrum an emission band spectrum corresponding to the absorption spectrum.[103] Here, he remarked, was a fresh example of multiple spectra. It now seemed clear to him that just as an elementary body might have two allotropic states, so it might have two spectra.

While today the halogens are not regarded as having allotropes they apparently were thought to have such, at least during the 1870s.[104] At that time "the question, still so obscure, of the allotropy of simple bodies" remained to be elucidated.[105] It was only in 1873 that Sir Benjamin Brodie, Professor of Chemistry at Oxford, experimentally deduced the chemical molecular form O_3 for ozone.[106] And no molecular forms had yet been suggested for the allotropes of other bodies.

In 1873, Salet published a lengthy résumé of his investigations on the spectra of the metalloids.[107] He had now found that bromine displays a line spectrum and an absorption spectrum which are quite different, and this, together with the similar iodine phenomena, seemed to him to put the question of double spectra beyond doubt.[108] Salet also added that in addition to sulphur and selenium, tellurium gives in the flame or electric discharge a band spectrum which is quite different from the line spectrum.

As we saw, Salet's results were readily accepted by Schuster, and indeed, they seem never to have been disputed by anyone. Even Ångstrom admitted, before his death in 1874, that elements such as iodine and sulphur could have more than one spectrum.[109] The important point

---

102 G. Salet, "Sur les spectres du sélénium et du tellure," *Comptes Rendus*, 73 (1871): 742–45.

103 G. Salet, "Sur le spectre primaire de l'iode," *Comptes Rendus*, 75 (1872); 76–77; in *Phil. Mag.*, 44 (1872): 156.

104 Schuster, for example, wrote that "we are aware that bodies, as iodine and sulphur, can give two spectra, but then the band spectrum is due to an allotropic state, which, from a spectroscopic point of view, behaves like a compound body." "Spectra of metalloids," *Nature*, 15 (1877): 447–48; p. 447.

105 Ed. Becquerel, "Report on the researches of M. Arn. Thenard concerning the actions of electric discharges upon gases and vapours," *Comptes Rendus*, 75 (1872): 1735–37; in *Phil. Mag.*, 45 (1873): 154–56.

106 Sir B. C. Brodie, "An experimental inquiry on the action of electricity on oxygen," *Phil. Trans.*, 162 (1872): 435–84; p. 484.

107 G. Salet, "Sur les spectres des métalloids," *Annales de Chimie et de Physique*, 28 (1873): 5–71.

108 *Ibid.*, pp. 49–52, and 70.

109 A. Ångstrom and R. Thalén, "Recherches sur les spectres des métalloids," *Nova Acta Regiae Societatis Scientiarum Upsaliensis*, 9 (1875), Article 9, p. 5. This was published by Thalén, Ångstrom's assistant and successor, after Ångstrom's death.

would seem to be that all of the elements in question were regarded as having allotropic forms. Each spectrum would then still correspond to a physically distinct entity, and the one-substance-one-spectrum principle could thus be preserved.

Ångstrom said that "in general all simple bodies showing the property of allotropy ought to give different spectra in the incandescent state provided that the said property occurs not only in the gaseous state but also at the temperature of incandescence." Thus oxygen, for example, ought to present two different spectra—"one for ordinary oxygen and the other for ozone." But as ozone is decomposed below the temperature of incandescence, there exists for oxygen only a single spectrum.

It is most important to note that the physicist Ångstrom, unlike a chemist, does not conceive of, say, individual oxygen atoms playing a role, or even of the existence of diatomic oxygen molecules. These were concepts which were yet to be developed by the physicist. Ångstrom merely speaks of allotropic states about which he is not too explicit. They are caused either by the molecules combining with one another or by the molecules taking up a different arrangement. With regard to spectral appearances, one of the allotropic forms of a body behaves essentially as a compound body.

Besides allotropy, another important factor in the ready acceptance of Salet's results was the design of his spectral tubes. This greatly reduced the likelihood of the elements investigated being contaminated. Salet also employed these tubes in studying the spectral properties of the gases hydrogen, oxygen, and nitrogen.

Salet says that although its lines may change in appearance, hydrogen nevertheless has only the one spectrum. He has verified Ångstrom's conclusions that the other spectra attributed to hydrogen are really those of acetylene and sulphur.[110] As for oxygen, Salet has been unable to reproduce the band spectrum and second line spectrum described by Wüllner. And he points out that, as Ångstrom has observed, Wüllner's additional oxygen spectra correspond very closely, one to the spectrum of carbon monoxide, the other to the spectrum of chlorine.[111]

Regarding nitrogen, Salet was well aware that it was much easier to prepare oxygen free of nitrogen than nitrogen free of oxygen. He was also aware that the beautiful band spectrum of nitrogen tubes was usually ascribed to an oxide of nitrogen. So he cleverly decided to investigate whether the nitrogen band spectrum could be obtained with pure ammonia gas. If so, then the question would be decided in favor of an

[110] G. Salet, "Sur les spectres des métalloïds," pp. 17–24.
[111] Ibid., pp. 35–36.

"elementary body" having a possible multiplicity of spectra. Salet therefore prepared a sheathed ammonia tube. The gas was passed over potassium and through the tube for a considerable time, and before disconnecting from the ammonia source the tube was heated near to red heat. Then the gas in the tube was rarified and excited electrically, whereupon the band spectrum of nitrogen appeared, together with the hydrogen spectrum. To Salet the question was resolved.[112]

This was, however, but one experiment. There was also Schuster's nitrogen experiment of the previous year (1872), in which the effects of oxygen had also been eliminated but which had nevertheless led to the opposite conclusion. Yet Schuster's experiment and result had not gone uncriticized.

The result was first criticized by Wüllner in 1872, and to understand the grounds of his criticism a slight digression is necessary. But what has to be said will also be important for the next section. Wüllner had at first admitted with Plücker and Hittorf that the band spectrum of an elementary gas appears at a lower temperature than the line spectrum. This view seemed to be confirmed by the fact that at relatively high pressures the bright lines also appeared simultaneously with the continuous spectrum, for the latter evidently denoted a high temperature. Yet, on the other hand, Wüllner had also observed that with gaseous carbon compounds the appearance of the line spectra is preceded by an extension of the band spectra, which seemed to indicate a lowering of temperature. He therefore conducted a series of experiments in which a thermocouple was inserted in the capillary part of a Plücker tube, and found that there was no relation between the temperature of the gas and the spectrum produced.[113] Thus the cause of the different spectra had to be sought elsewhere, and Wüllner found it in the nature of the discharge. From many observations he concluded that line spectra are produced only by the discontinuous or spark discharge, whereas the continuous discharge invariably produces band spectra.

Applying these conclusions to Schuster's experiment, Wüllner says he cannot see in it a demonstration against the occurrence of double spectra—simply because Schuster has employed the spark discharge throughout.[114] Wüllner was soon afterward to explain that one could object to his objection by saying that in perfectly pure nitrogen gas the discharge can only occur in the form of a spark, as is the case with the vapors of bromine and iodine (this was before he knew of Salet's

112 *Ibid.*, p. 52.
113 A. Wüllner, "Ueber die Spectra der Gase in Geissler'schen Röhren; die Entstehung der Spectra verscheidener Ordnung," *Ann. der Physik,* 147 (1872): 321–53.
114 *Ibid.*, p. 353.

work).[115] However, in experimenting with pure nitrogen Wüllner obtained results entirely in accordance with those previously obtained with air, which had shown nitrogen to have two spectra depending upon the nature of the discharge.[116]

Schuster's work was questioned in another way by C. H. Stearn of Liverpool. In a letter to the journal *Nature*, Stearn related that he and a friend named Lee had repeated Schuster's experiment without, however, having been able to confirm Schuster's result.[117] This, we recall, was that the line spectrum of nitrogen could be obtained under all conditions of temperature and pressure when all of the oxygen had been removed from the nitrogen by the sodium. Stearn and Lee found that the line spectrum of nitrogen could be obtained at all pressures, but Stearn adds: "That it can be obtained at all temperatures, by which, I presume, Mr. Schuster means either with or without the Leyden jar, is certainly contrary to our experience."

Furthermore, upon heating the sodium Stearn and Lee invariably found that an increase of pressure occurred due to the liberation of hydrogen, presumably from water vapor contained in the gas. Having exhausted the tube a second time they obtained a spectrum of lines with the "simple" current. This, however, was not the nitrogen spectrum but the "second so-called hydrogen spectrum first described by Plücker." Upon introducing a Leyden jar into the circuit this spectrum was replaced by the ordinary hydrogen spectrum.

Stearn was also struck by the fact that only a few of the lines given by Schuster coincided with those of nitrogen, while many of nitrogen's most brilliant lines, including the characteristic double green line, were not represented in Schuster's spectrum.

Within two months Schuster replied from Heidelberg.[118] He had re-examined the spectrum in question and had found that it was indeed not that of nitrogen. While he was unable to say to what it was due, he nevertheless did not think that its formation could be put forward as proof that the band spectrum is not due to oxides of nitrogen. He promised to repeat his experiment at the earliest opportunity, and hoped then to bring the question to a satisfactory conclusion.

Thus the result of Schuster's experiment was being called in ques-

115 A. Wüllner, "Ueber die Spectra der Gase in Geissler'schen Röhren," *Ann. der Physik*, 149 (1872): 103–12.
116 A. Wüllner, "Ueber die Spectra der Gase in Geissler'schen Röhren; die Entstehung der Spectra verscheidener Ordnung," *Ann. der Physik*, 147 (1872): 321–353; pp. 325–29.
117 C. H. Stearn, "Spectrum of nitrogen," *Nature*, 7 (1873): 463.
118 A. Schuster, "Spectrum of nitrogen," *Nature*, 8 (1873): 161.

tion about the time when Salet described his experiment pointing to the opposite conclusion. Salet and Wüllner now agreed that nitrogen had two characteristic spectra. But Ångstrom and presumably also Schuster were still opposed to such a view. In 1874 Ångstrom and Thalén concluded that the band spectrum of nitrogen tubes was due to nitrogen dioxide.[119]

Arrived at this impasse we can well imagine the proponents of double spectra seeking an indisputable demonstration of their viewpoint. And doubtless this was the motivation which led in 1875 to Salet's telling discovery that the electric spark when passed through a mixture of nitrogen and oxygen synthesizes nitrogen peroxide—whose spectrum was well known and does not coincide at all with the band spectrum produced by a nitrogen tube![120] Later in the same year Salet pointed out, no doubt with Stearn and Wüllner in mind, that the results of Schuster's experiment had been sharply criticized and finally rejected by Schuster himself.[121] Nevertheless, Salet was still not satisfied, and so repeated Schuster's experiment by himself.

In the following year he reported that heated or even volatilized sodium was not always sufficient to cause the nitrogen bands to disappear, and that when they did disappear they were replaced by several spectra, not one of which was that of nitrogen.[122] Further investigation indicated that the effect of the sodium was probably to absorb the nitrogen. So Salet attached a manometer to the spectral tube and indeed found that as the sodium changed appearance the pressure within the tube fell to zero. Then when the resultant sodium product was placed in water it gave ammonia, from which Salet deduced that it was sodium nitride.[123]

Thus the disappearance of the nitrogen band spectrum was to be explained by the sodium absorbing not an oxygen impurity, but the nitrogen itself. This absorption, however, could be prevented if care was taken to keep the volatilized sodium away from the electrodes; and in this case the nitrogen band or line spectrum could be produced at will.[124] It remained then to explain the origin of the lines described

119 A. Ångstrom and R. Thalén, "Recherches sur les spectres des métalloids," pp. 16–17.
120 G. Salet, "Sur les spectres multiple," *Association Française pour l'Advancement des Sciences, Comptes Rendus,* (1875): 485–89; p. 488.
121 G. Salet, "Sur les spectres doubles," *J. de Physique,* 4 (1875): 225–27.
122 G. Salet, "Sur le spectre d'azote et sur celui des métaux alcalins, dans les tubes de Geissler," *Comptes Rendus,* 82 (1876): 223–26; p. 224. And, *J. de Physique,* 5 (1876): 95–97; p. 95.
123 See second article of previous note, p. 97.
124 *Ibid.,* p. 96.

by Schuster, and by inspection Salet found that they were probably to be attributed to sodium vapor.[125]

For almost three years Schuster had written nothing on spectrum analysis, but in 1877 he published an abstract of the above-mentioned paper of Ångstrom and Thalén in *Nature*, and followed it two weeks later with a letter containing observations both on their views and on the spectra of nitrogen.[126] Whereas in connection with gases Schuster had previously been an advocate of the "one-element-one-spectrum" principle, he now showed himself to be an enthusiastic apostle of multiple spectra. This reversal is important, for with Ångstrom dead and Thalén apparently more interested in other aspects of spectroscopy there remained no active opponents of multiple spectra.

Schuster acknowledges the correctness of Salet's explanation of the nitrogen experiment and adds:

> If I have refrained hitherto from acknowledging the justice of Mr. Salet's conclusions, it is only due to the fact that I felt a natural curiosity to repeat his experiments; I have not yet been able to do so, but I have no doubt what the result would be. Mr. Salet's paper was only published after Professor Ångstrom's death, and I cannot help thinking that the professor would have considered his experiments as conclusive against the assumption that the bands of nitrogen are due to an oxide of nitrogen.[127]

Thus while Plücker and Hittorf discovered double spectra, and Wüllner defended and furthered the idea, it was Salet who finally succeeded in placing the existence of multiple spectra beyond doubt.

In conclusion Schuster challenged any remaining skeptics in the following confident terms:

> If anyone still believes that an element can only have one spectrum at the temperature of the electric spark, I propose to him the following problem: Let him take the three gases, carbonic acid, acetylene and oxygen. If he investigates their spectra carefully he will at least find ten different spectra (two of which I only discovered lately). Out of carbonic acid alone he can obtain six. Let him find a sufficient number of possible compounds to account for all these spectra.[128]

As may be seen from Schuster's next publication in 1879 the two recently discovered spectra probably belonged to that gas which the champions of multiple spectra—Plücker, Hittorf, Wüllner, and Salet—had

[125] *Ibid.*, p. 97; and, Salet, "Sur le spectre d'azote," *Comptes Rendus*, p. 224.
[126] A. Schuster, "Researches on the spectra of metalloids," *Nature*, 15 (1877): 401–2. "Spectra of metalloids," *Nature*, 15 (1877): 447–48.
[127] *Ibid.*, p. 448.
[128] *Ibid.*

found either unresponsive or difficult, namely, oxygen. In all, Schuster distinguishes four oxygen spectra: one band, one continuous, and two line spectra.[129]

In the same year in which these results were published the British Association for the Advancement of Science established a committee to report upon the current state of knowledge in spectrum analysis. Schuster's name does not appear among those of the committee at its inception, but, when it reported in 1880, two of the four sections of the report were written by Schuster. Of multiple spectra he says: "Plücker's discovery . . . was established in the case of all metalloids which have been sufficiently well studied. There is now among those best able to judge a general agreement on the facts, although great differences exist as to their interpretation."[130] It is to this latter subject that we must now turn.

4. Explanations of Multiple Spectra and the Atomic-Molecular Theory of Spectra

Two theories were advanced to explain multiple spectra, the first in 1872. At this time the second was beginning to take form in that, as we have seen, it was being suggested that each allotropic state of an element gives a characteristic spectrum. During the remainder of the decade the further development of this second theory was to carry physical thought on matter through a considerable evolution. The emerging physical view came together with the chemical view, but for many spectroscopists their complete union was short-lived. The outcome, however, was that the physical view of matter had been forced into sharper focus.

The earlier of the two theories was given by Wüllner. In the previous section it was noted that he had been led to conclude, contrary to the accepted view, that the species of spectrum given by a rarified gas in a discharge tube did not depend on the temperature of the gas. Rather, the spectrum was determined by the nature of the electrical discharge. By means of experiments on air, hydrogen, and oxygen, Wüllner had become convinced that line spectra are given by the high tension discontinuous or spark discharge, whereas band spectra are given by the low tension continuous discharge.

To explain these facts Wüllner made use of a relationship established two years earlier by Friedrich Zöllner in a theoretical consideration,

---

129 A. Schuster, "On the spectra of metalloids—Spectrum of Oxygen," *Proc. Roy. Soc.*, 27 (1878): 383–88; and *Phil. Trans.*, 170 (1879): 37–54.
130 A. Schuster, "Spectra of metalloids," *Rep. British Assn.*, 48 (1880): 258–74; p. 258.

based on Kirchhoff's law of emission and absorption, of the influence of temperature and pressure on the spectra of incandescent gases.[131] Among other things, Zöllner had demonstrated that for a given temperature the spectrum of a gas ought to depend essentially on the thickness or density of the layer of luminous gas. A diminution in the thickness or density of this layer should reduce the spectrum to a number of bright lines corresponding to those wavelengths for which the gas has the greatest emissive power.

Accepting these conclusions, Wüllner maintains that variations in the thickness or density of the luminous layer are the principal causes of the production of spectra of different orders in Geissler tubes.[132] In the spark discharge only a thin thread of gaseous molecules along the path of the discharge becomes luminous. Hence, a line spectrum results. On the other hand the continuous discharge "more or less" fills the entire section of the tube and light is therefore emitted by a considerably greater thickness of gas. All of the wavelengths which the gas is capable of emitting at the particular temperature should be given. The result is then a band spectrum, or so Wüllner maintains.

This explanation of double spectra was given at a time when Wüllner was their sole defender. To others the order of priority was first to establish the existence or nonexistence of double spectra and only then to attempt their explanation—if required. Thus we find no comment on Wüllner's theory until 1875, when it was attacked by a fellow countryman who was to become a much respected student of discharge-tube phenomena, Eugene Goldstein.

Goldstein employed a clever rotating mirror device which enabled him to distinguish clearly between continuous and discontinuous discharges.[133] Contrary to Wüllner's observations, Goldstein found that the continuous discharge could give line as well as band spectra, and the discontinuous discharge band as well as line spectra. And so he concluded that the mode of discharge was not a determining factor as regards the nature of the spectrum. Furthermore, Goldstein could not confirm that in the spark discharge only a few molecules, and therefore only a thin layer of gas, become luminous. And he completely undermined Wüllner and Zöllner's position by showing that band spectra may be produced by gas in a capillary tube, while line spectra may be obtained from much thicker layers of gas. Goldstein concluded from his

[131] F. Zöllner, "Ueber den Einfluss der Dichtigkeit und Temperatur auf die Spectra glühender Gase," *Ann. der Physik*, 142 (1871): 88–111; in *Phil. Mag.*, 41 (1871): 190–205.
[132] A. Wüllner, Über die Spectra der Gase in Geissler'schen Röhren," *Ann. der Physik*, 147 (1872): 321–53; pp. 340–48.
[133] E. Goldstein, "Über Beobachtungen an Gasspektris," *Ann. der Physik*, 154 (1875): 128–49; in *Phil. Mag.*, 49 (1875): 333–45.

experiments that any required spectrum could be obtained at any pressure, however small, provided the gas be at the required temperature. But he attempted no theoretical explanation.

Wüllner conceded that Goldstein's experiments were "very interesting," but refused to see in them any contradiction of his theory. "I see, on the contrary, in the experiments of Mr. Goldstein in general a corroboration of my view, which requires a band spectrum whenever extensive masses of gas are rendered luminous."[134]

This reply is typical of Wüllner's responses to the various criticisms made by several other prominent spectroscopists and again by Goldstein over the next twenty years.[135] Some of the criticism was off the mark, but on the whole it was sound.[136] As in establishing the existence of double spectra, so here with their explanation, Wüllner was to find himself alone. But whereas with the passage of time he was joined in the former instance by all other spectroscopists, in the latter his position was to remain an isolated one. It would be interesting to know the psychology of the individual scientist who opposes the majority opinion of his colleagues, and especially the psychology of Wüllner who resisted the opposition both when he was right and when he was wrong. But it would serve no useful purpose to consider in detail the various objections and Wüllner's replies to them. A few general remarks will suffice.

Heinrich Kayser was perhaps Wüllner's last critic, when in the early 1890s it was still thought worth the effort to attack Wüllner's theory.[137] About ten years later when looking back from the perspective of 1902 Kayser wrote:

Wüllner's views were developed by him in numerous papers, attacked by others and upheld by him. In spite of these and in spite of the presentation in Wüllner's *Lehrbuch der Physik* I must admit that his views were not wholly

---

134 A. Wüllner, "Einige Bemerkungen zu Herr Goldstein's 'Beobachtungen an Gasspektris,'" *Monatsberichte Berlin Akad.* (1874): 755–61; in *Phil. Mag.,* 49 (1875): 448–53; p. 449.

135 E. Goldstein, "Ueber das Bandenspectrum der Luft," *Ann. der Physik,* 15 (1882): 280–88; p. 288. A Wüllner, "Einige Bemerkungen zu den Mitteilungen der Herren Hasselberg und Goldstein," *Ann. der Physik,* 17 (1882): 587–92.

Hermann Ebert, "Ueber den Einfluss der Dicke und Helligkeit strahlenden Schicht auf das Aussehen des Spectrums," *Ann. der Physik,* 33 (1888): 155–58. For Wüllner's reply see *Ann der Physik,* 34 (1888): 647–61.

E. Wiedemann, "Ueber das thermische und optische Verhalten von Gasen unter dem Einflusse electrischer Entladung," *Ann. der Physik,* 10 (1880): 202–57; in *Phil. Mag.,* 10 (1880): 357–80, 407–27; pp. 426–27. H. Kayser, "Ueber den Ursprung des Banden- und Linienspectrums," *Ann. der Physik,* 42 (1891): 310–19.

136 See E. Wiedemann, "Ueber das thermische und optische Verhalten von Gasen," *Phil. Mag.,* p. 426, where he acknowledges a previous criticism to have been shown invalid.

137 H. Kayser, "Ueber den Ursprung des Banden- und Linienspectrums."

clear to me; for the observations which Wüllner communicated often seem to me to contradict his words and I do not understand how one can bring both into harmony.[138]

Goldstein would have agreed.

One powerful objection was that if thick layers of gas give only band spectra, how is it that the enormously thick layers met with in solar and stellar phenomena give line spectra? This difficulty was posed by Norman Lockyer with whom the second, or molecular, theory of multiple spectra originated in 1874.[139] I put it this way because at the time Lockyer was not directly concerned with multiple spectra.

In order to understand the course of Lockyer's thought it is necessary to digress a little and consider certain results at which he had arrived in 1873. These concern the first suggestion of his dissociation hypothesis which is closely related to the molecular theory of multiple spectra and which will be considered in the following section.

Seven years of intensive study of the solar and stellar spectra had revealed to Lockyer certain suggestive facts, of which it is sufficient to paraphrase only the more directly relevant ones.

1. Spectra of the metalloids and their metallic compounds are distinguished from metallic spectra by their banded structure.
2. In all probability there are no compounds ordinarily present in the sun's reversing layer.
3. When a metallic compound is dissociated by the spark discharge the band spectrum dies out and the elemental lines appear.
4. According to its spectrum the sun may be regarded as belonging to class β of stars intermediate between class α with much simpler spectra of the same kind and class γ with much more complex spectra of a different kind.
5. Sirius, belonging to class γ is "the brightest (and therefore hottest?) star in our northern sky." Hydrogen is the only element which is known with certainty to be present in Sirius, the other metallic lines being exceedingly thin and thus indicating a small proportion of metallic vapors. The hydrogen lines in this star are *"enormously distended"* showing that the chromosphere is largely composed of that element.
6. Examples of class γ are the red stars whose spectra are composed of channelled spaces and bands. Hence, the reversing layers of these stars probably contain metalloids, or compounds, or both, in great

138 H. Kayser, *Handbuch der Spectroscopie* (Leipzig, 1902), 2: 244.
139 Lockyer's objection is mentioned by Wiedemann, "Ueber das thermische und optische Verhalten von Gasen," *Phil. Mag.,* p. 426.

quantity; and not only is hydrogen absent, but the metallic lines are reduced in thickness and intensity, which in light of (3) "may indicate that the metallic vapours are being *associated*. It is fair to assume that these stars are of a lower temperature than our sun." Contemplation of these observations led Lockyer to ask whether they could not

be grouped together in a working hypothesis which assumes that in the reversing layer of the sun and stars various degrees of "celestial dissociation" are at work, which dissociation prevents the coming together of atoms which, at the temperature of the earth and at all artificial temperatures yet attained here, compose the metals, the metalloids, and compounds.

On this working hypothesis, the so-called elements not present in the reversing layer of a star will be in course of formation in the coronal atmosphere and in course of destruction as their vapour densities carry them down.[140]

I have already indicated in section 1 of this chapter that during the nineteenth century there was a widespread belief that all matter is composed of the one kind of ultimate atom. This idea will be considered at some length in the following section.

At the time when Lockyer wrote, the metals could be distinguished spectroscopically from the metalloids in that, as he expressed it, the emission and absorption spectra of the former were identical, whereas in the case of the latter they were different. That is, the metals appeared to have one, and the metalloids two, spectra. In passing from a lower to a higher temperature, says Lockyer, the "plasticity" of the metalloids comes into play and we get a new molecular arrangement, as in the case of ozone and oxygen.[141]

In the following year, 1874, however, Roscoe and Schuster found the vapors of sodium and potassium at low temperatures to have absorption spectra corresponding to those of the metalloids.[142] In the same year this result was incorporated by Lockyer in a paper in which he gave a preliminary account of research which led him to conclude that "a mass of elemental matter . . . is continually broken up as the temperature (including in this term the action of electricity) is raised."[143] This is a remarkable statement for, as pointed out earlier, the physicist had previously regarded all of the molecules of an elementary substance, excepting an allotropic substance, as being identical.

140 J. N. Lockyer, "Researches in spectrum-analysis in connection with the spectrum of the sun—No. III," *Phil. Trans.*, 164 (1874): 479–94; p. 492.
141 *Ibid.*
142 H. Roscoe and A. Schuster, "Note on the absorption-spectra of potassium and sodium at low temperatures," *Proc. Roy. Soc.*, 22 (1873–74): 362–64.
143 J. N. Lockyer, "Spectroscopic notes—No. II. On the evidence of variation in molecular structure," *Proc. Roy. Soc.*, 22 (1873–74): 372–74.

Starting from the fact that when a chemical compound is dissociated into its elements a change of spectrum occurs, Lockyer reverses and generalizes it to postulate that all changes from one spectrum to another are accompanied by a corresponding change in molecular complexity. In essence Lockyer is supporting one and advancing two other hypotheses, namely: that elementary bodies have multiple spectra; that elementary bodies can exist in different molecular forms; and that each molecular form has a characteristic spectrum. On the basis of experiment he arrives at the following series in which spectral appearances are listed against increasing molecular complexity.

Degree "of molecular complexity of molecule"	Spectrum
1st	Line spectrum
2nd	Channelled space spectrum
3rd	Continuous absorption at blue end not reaching to less refrangible end
4th	Continuous absorption at red end not reaching to more refrangible end
5th	Unique continuous absorption

Thus according to Lockyer an elementary body could have as many as five different molecules. But he had no way of determining just how many ultimate atoms were in each molecule.

In an article entitled "Atoms and Molecules Spectroscopically Considered," published in the same year in his own journal, *Nature,* Lockyer carefully distinguishes between chemical and physical uses of the words "atom" and "molecule." He quotes Edward Frankland's definition of the chemical atom which is similar to Berthelot's—an atom is the smallest proportion by weight in which an element enters into or is expelled from a chemical compound.

There is one passage in Lockyer's paper which deserves special attention:

. . . the spectroscope has to a great extent vindicated the theory [of radiation] stated by Professor Maxwell. The question is, has it taken us further? Perhaps not yet, but I think it will be found that what chemists picture to themselves as the atom, as contradistinguished from what they weigh, and physicists the molecule, is the particular atom, molecule, particle, or whatever name you choose to call it, with which high tension electricity gives us a spectrum of lines.

You recollect that I said that in many of the monad metals it was obtained in the first stage of temperature, in the case of the dyads and metalloids with higher stages. If the true atom be that which gives a line spectrum, many anomalies will fall to the ground. . . . . If you allow that in the line spectrum an atom is at work, in channelled spectra and continuous spectra molecular aggregations, you will see at once that Professor Maxwell and others will be able to get a much sharper definition of atom and molecule than they have now.[144]

Thus Lockyer now proposed to introduce the "true" or spectroscopic atom. This atom is equivalent to both the chemist's atom and the physicist's simplest molecule. Here we have the first suggestion of an atomic-molecular theory of spectra.

Chemists, who apart from Lecoq de Boisbaudran had not troubled themselves with explaining just how a Lucretian atom could produce a spectrum, could welcome Lockyer's view.[145] This latter was known to Salet when at the 1875 meeting of the Association Française pour L'Avancement des Sciences he proposed a theory of double spectra. Salet essentially agrees with Lockyer, although the fact that he is a chemist makes for one interesting difference.

As for the theory of double spectra, it is easy to imagine one. As heat has ordinarily the effect of simplifying chemical compounds and of reducing them to their elements; as, on the other hand, elementary channelled spectra are observed at the lowest temperatures and are very similar to those of compound bodies, one can suppose that they are produced by aggregations of homogeneous atoms. This assumption will appear natural above all to chemists, for whom for a long time chlorine has been the chloride of chlorine (cl^2), sulphur, the sulphide of sulphur (S^2), etc. But these molecules are themselves capable of being resolved into atoms: it is to this *allotropy of elevated temperatures* that the variation of spectra seems due.—Indeed, for metals for which the molecule coincides with the atom no channelled spectra have been reported. This theory evidently calls for experimental verification; but, true or false, it is not useless, since it is made to work.[146]

Thus Salet ascribes line spectra to chemical atoms and band spectra to chemical diatomic molecules. This is a step beyond Lockyer.

As further support for his theory Salet mentions in a footnote that a certain acoustical experiment interpreted in terms of kinetic theory has recently shown mercury vapor to be monatomic. And this vapor gives a line spectrum. We shall consider the experiment and its interpretation in chapter IV.[147] For the present it is sufficient to say that

144 J. N. Lockyer, "Atoms and molecules spectroscopically considered," *Nature,* 10 (1874): 69–71, 89–90; p. 70.
145 For Lecoq see pp. 199–200.
146 G. Salet, "Sur les spectres multiples," p. 489.
147 See p. 161.

like Salet some physicists readily accepted the result, while others did so reluctantly.

Salet had brought chemical ideas into physics with the help of the idea of allotropy. In addition to not being well understood, "allotropy" had a different meaning in the 1870s from that which it has today. It was then much more general and included substances having diatomic molecules, as can be seen from the above quotation. With this conception of allotropy the chemist could offer the physicist an acceptable explanation of the double spectra of such substances as hydrogen and nitrogen—bodies which he previously regarded as having no allotropes.

It is of interest that elsewhere in the same year Salet gives allotropy a somewhat different meaning in explaining that "very serious chemical arguments lead one to think that the molecule of simple bodies is often composed of several similar atoms, and that different groupings of these atoms give rise to different allotropic states of elements."[148] Here he seems to be saying that an elementary body has but one molecule from the standpoint that it is composed of a definite number of atoms, but that these atoms can take on different arrangements and so give rise to several allotropes. In this way Salet could have accounted for multiple spectra without departing from the chemical view. We shall shortly see, however, that Schuster and others could not reconcile multiple spectra with the chemical view.

For some time chemists had assumed that elementary bodies underwent a dissociation under chemical forces. Now in the mid-seventies spectroscopy was providing indirect physical evidence that heat and electricity also brought about a dissociation. It was not until 1878 that Victor Meyer, on the basis of vapor density studies, began to provide direct physical evidence for thermal dissociation.[149]

The outcome of these various developments was that most physicists now adopted the chemists' ideas of atoms and molecules. For these physicists a molecule was no longer an aggregate of an unknown number of ultimate atoms but rather a combination of a definite number, usually two, of chemical atoms. The chemical atom was generally regarded as being composed of a definite, although unknown, number of ultimate atoms—the number being different for each element. This latter will be seen in the following section.

Our consideration of Salet's work perhaps helps to explain why

---

148 G. Salet, "Sur les spectres doubles," p. 226.
149 See J. R. Partington, *A Short History of Chemistry* (3d ed.; New York, 1957), p. 313.

Hermann von Helmholtz, as reported by James Moser in 1877, expressed views similar to Salet's.[150] Upon hearing of Helmholtz's opinion Schuster remarked that, this "is precisely the view first put forward by Lockyer, and it has thus received a striking confirmation from an independent quarter."[151] But we know that Lockyer's original view was not nearly so definite as that now expressed. In the following year (1878) another German physicist, Eilhard Wiedemann, espoused an identical view.[152] Wiedemann's atomic and molecular models are described in detail in chapter IV.[153]

Within a year, however, this united chemical-physical view was in trouble. Schuster had found oxygen to give two distinct line spectra.[154] Clearly there could not be different oxygen atoms for each of these, but there could be different oxygen molecules for each. Thus in a sense we find a return to a molecular theory of spectra. But it is not the former physical molecular theory, for now each of the unspecified number of atoms within a molecule is a chemical and not an ultimate atom. And furthermore, the chemical atoms vibrate internally as well as with respect to one another within the molecule.[155] Essentially we have an alternative atomic-molecular theory of spectra.

Of the oxygen line spectra, Schuster says that

one appears under precisely the same circumstances as the band spectrum of nitrogen, and it changes into the other line spectrum almost identically in the same way as the band spectrum of nitrogen changes into the line spectrum. The homogeneousness of the phenomena seen in vacuum tubes is greatly increased, if we consider the one line spectrum of oxygen as the representative of the band spectrum of nitrogen, and I shall therefore consider it as such. I have called the spectrum the compound line spectrum of oxygen, because, according to the theory of molecular combination, it would be due to a more complicated molecular structure than the other (the elementary) line spectrum. I must, however, add that I do not at all feel certain that the two line spectra are necessarily to be explained in the same way as the line and band spectra. I have found it necessary, for the sake of clearness, to explain the results of the investigation in the language of a definite theory.[156]

150 J. Moser, "Die Spectren der chemischen Verbindungen," *Ann. der Physik,* 160 (1877): 177–99.

151 A. Schuster, "The spectra of chemical compounds," *Nature,* 16 (1877): 193–94.

152 E. Wiedemann, "Untersuchungen uber die Natur der Spectra," *Ann. der Physik,* 5 (1878): 500–524.

153 See p. 179.

154 A. Schuster, "On the spectra of metalloids. Spectrum of oxygen," *Phil. Trans.,* 170 (1879): 37–54.

155 A. Schuster, "The teachings of modern spectroscopy," *Popular Sc. Monthly,* 19 (1881): 466–82. This paper is discussed in part in chapter IV, p. 171.

156 A. Schuster, "On the spectra of metalloids. Spectrum of oxygen," pp. 40–41.

The last two sentences call for some explanation. Schuster lacked conviction not because of the Salet–Helmholtz view, but rather because of what he calls the molecular collision theory, which he regards as an alternative to the molecular combination theory of double spectra. But in this last respect Schuster is wrong, for no one attempted to explain double or multiple spectra on the molecular collision theory. And in the following year (1880), Schuster cited Wüllner's theory as the only rival to the theory of molecular combination.[157] As we shall see in the following section, the theory of molecular collisions was an alternative to Lockyer's dissociation hypothesis.

Turning to continuous spectra Schuster says:

> that the discontinuous spectra of different orders (line and band spectra) are due to different molecular combinations, I consider to be pretty well established, and analogy has led me (and Mr. Lockyer before me) to explain the continuous spectra by the same cause; for the change of the continuous spectra to the line or band spectrum takes place in exactly the same way as the change of spectra of different orders into each other.[158]

Negative glow spectra are also to be explained by assuming a particular molecule—"it will be found simpler to assume that a definite molecular combination is formed at the negative pole than to suppose that exterior forces peculiar to the pole modify the period of vibration."[159]

Thus we are back to Lockyer's initial position with an elementary substance having as many different molecules as it has different spectra. Only this time the molecules are composed of chemical atoms which vibrate internally. But there is just as much vagueness about the precise nature of the molecules. For example, just how do the molecules producing line spectra differ from the other molecules?

There were many physicists, however, who continued to work with the chemical view of the atom giving rise to the line spectrum and the diatomic molecule to the band spectrum. This was sufficient when only one line spectrum was known for a substance. But Schuster's view was the more general and therefore the better alternative around 1880. Later the embarrassment of the chemical view when faced with two line spectra would disappear with the development of the charged atom. For then physicists like J. J. Thomson could ascribe different line spectra to positively and negatively charged atoms. Nevertheless, Schuster's view continued to be adopted by many, as we shall see in

[157] A. Schuster, "On our knowledge of spectrum analysis," *Rep. British Assn.*, 48 (1880): 258–98; p. 285.
[158] A. Schuster, "On the spectra of metalloids. Spectrum of oxygen," p. 39.
[159] *Ibid.*, p. 41.

later chapters. Indeed it was still favored by some in the first decade of this century.

But in spite of these differences, all were united in accepting an atomic-molecular, rather than Wüllner's, explanation of multiple spectra.

5. Lockyer's Hypothesis of the Dissociation of the Elements

Lockyer's hypothesis of the dissociation of the elements is closely related to the atomic-molecular theories of multiple spectra. All were based on analogy with the spectral behavior of compound molecules, and the hypothesis is really an extension of the theories. For if we imagine passing molecule by molecule from the most complex molecule producing the continuous spectrum to Schuster's least complex molecule and the chemical atom, then Lockyer's hypothesis carries us one step beyond. The chemical atom becomes dissociated. Here we are dealing with that view of matter mentioned in the first and previous sections as generally favored by physicists. The chemists' elemental atom is regarded as a 'molecular' aggregate of ultimate atoms.

We saw how, during the course of his solar spectral work, the dissociation hypothesis was suggested to Lockyer. In continuing to map the solar spectrum Lockyer's attention was drawn in 1874 to another suggestive fact. He noticed that the two solar lines H, belonging to calcium, were each considerably wider than in a normal calcium spectrum.[160] Lockyer was thus led to study the behavior of the blue end of the calcium spectrum at various temperatures, as shown in stages one to four of the accompanying plate (Fig. 2, from Lockyer's Studies in Spectrum Analysis, page 191). At stage one the spectrum of undissociated calcium chloride, having no lines in the blue, is given. At stage two some of the chloride molecules are dissociated by the induced current and a calcium line is seen. Next, a weak arc current produces the calcium "spectrum." The single line to the right is much broader than the two lines to the left, and it becomes reversed. Finally, a stronger current produces stage four, where the three lines are almost equally wide and all are reversed.

Lockyer considers stage three in light of the spectral behavior of a compound where the band spectrum "is reduced as dissociation works its way, and the spectrum of each constituent element makes its appearance." If in stage three "we take the wide line as representing the banded spectrum of the compound, and the thinner ones as represent-

---

[160] J. N. Lockyer, Studies in Spectrum Analysis (New York, 1878) (referred to below as SSA), p. 190.

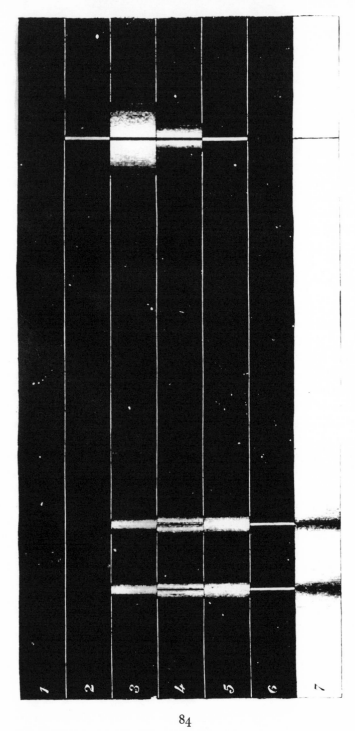

Fig. 2. Blue End of the Calcium Spectrum at Various Temperatures

ing the longest elemental lines making their appearance as the result of partial dissociation, we have, by hypothesis, an element behaving like a compound."[161]

Lockyer next reasons that if his hypothesis be true, then at lower temperatures the single line ought to be much brighter, and at higher temperatures the double lines. So he experimented first with a large induction coil and small Leyden jar (stage 5) and then with a large coil and large jar (stage 6). In the latter case only the double lines appear, in contrast to stage two, where we have only the single line of calcium. So Lockyer assumes that by analogy he is dealing exclusively with dissociated calcium at stage six.[162]

George Gabriel Stokes, however, could not accept this conclusion, and in a letter to Lockyer argued that:

When a solid body such as a platinum wire, traversed by a voltaic current, is heated to incandescence, we know that as the temperature increases, not only does the radiation of each particular refrangibility absolutely increase, but the properties of the radiations of the different refrangibilities is changed, the proportion of the higher to the lower increasing with the temperature. . . . Hence, while I regard the facts you mention as evidence of . . . high temperature . . . , I do not regard them as *conclusive* evidence of the dissociation of the molecules of calcium.[163]

To this Lockyer replied that it was the less, and not the more, refrangible calcium line which increased in intensity with elevation of temperature. And he asked Stokes: "Are you quite sure that the molecular structure of the platinum wire is constant while it behaves as you say it does?"[164]

Lockyer now sought further evidence for his hypothesis in stellar spectra. To him it was "abundantly clear that if the so-called elements, or more properly speaking their finest atoms—those that give us line spectra—are really compounds, the compounds must have been formed at a very high temperature."[165] But it was easy to imagine even higher temperatures capable of dissociating these "combinations" into "atoms;" and of this dissociation "there may be no end."

Lockyer continues that

we are justified in supposing that our calcium, once formed, is a distinct entity, whether it be an element or not, and therefore, by working at it alone, we shall

161 *Ibid.*, p. 192.
162 *Ibid.* pp. 192–93. See also J. N. Lockyer, "Preliminary note on the compound nature of the line spectra of elementary bodies," *Proc. Roy. Soc.*, 24 (1876): 352.
163 Published in *ibid.*, and in *SSA*, pp. 194–95.
164 Published in *ibid.*, and in *SSA*, p. 195.
165 *Ibid.*, p. 196.

never know, even if its dissociation be granted in the future, whether the temperature produces a simpler form or more atomic condition of the same thing, or whether we actually break it up into $X + Y$, because neither X nor Y will ever vary.

But if calcium be a product of a condition of relatively lower temperature, then in the stars, hot enough to enable its constituents to exist uncompounded, we may expect these constituents to vary in quantity; there may be more of X in one star and more of Y in another; and if this be so, then the H and K lines will vary in thickness, and the extremest limit of variation will be that we shall only have H representing, say X in one star, and only have K representing say Y in another. Intermediately between these extreme conditions of cases, we may have cases in which, though both H and K are visible, H is thicker in some and K is thicker in others.[166]

In 1878 there was but a single piece of evidence tending to support such a view. The astronomer William Huggins had noted that only one of the solar H lines was represented in the spectrum of a-Lyrae.[167]

However, in continuing his mapping of the solar spectrum Lockyer made an important discovery which led to a major development of the dissociation hypothesis. In previous mapping work Kirchhoff, Ångstrom, and Thalén had found many metallic spectra to contain common lines.[168] But there was no means of deciding whether these coincidences were accidental or due to impurities of one metal in another. None, that is, before Lockyer devised his method of long and short lines. It was in applying this method in the examination of coincident lines that Lockyer made the discovery spoken of.

To understand the method of long and short lines let us imagine an electric arc produced between poles lying along a horizontal line.[169] By means of a lens an image of the arc is projected on to the vertical slit of a spectroscope. A thin section of the arc light may then be examined. The arc light is supposed hotter at its core and cooler toward its edges, thus some differences would be expected in the spectral appearances of the two positions. And this is found to be the case. The spectrum consists of long and short lines as shown in Figure 3. The long lines are produced at every point of the arc light, whereas the short lines are only produced at or near to the core. Each element displays long and short lines.

Lockyer is unable to explain the difference between the two sets of lines, but he does know that as the current through the arc is reduced the short lines are first to disappear. This is the important point. The

---

166 *Ibid.*, p. 197.
167 Mentioned by Lockyer, *ibid.*, p. 196.
168 See, for example, G. Kirchhoff, *Researches on the Solar Spectrum*, 1:10.
169 The method is discussed in *SSA*, and in Lockyer, *Chemistry of the Sun* (London, 1887).

Fig. 3. Illustration of Long and Short Lines in a Spectrum

long lines are first to be produced and are therefore most characteristic of the element. But that is not to say that they are necessarily the brightest lines, although sometimes they are.

When Lockyer, in employing his method, refers to the long or short lines of a spectrum he does not mean that they literally appear long or short in that spectrum. This is where the terminology causes some confusion. What is meant is that the spectrum was previously examined as an arc or other convenient spectrum, when its lines did literally appear either long or short. The wavelength of a long line, say, was then determined, and afterward when a line of that wavelength was seen in the spectrum produced under any conditions it was referred to as a "long" line.

To see how the method of long and short lines is employed in considering lines common to two or more metallic spectra, consider the two spectra A and B. If it is found that the common line λ is the "longest" line of A, say, then we eliminate λ from B. We proceed to the second, third, etc., next "longest" lines of A, and if they are found in B we likewise eliminate them. We continue in this manner until a line of A is reached which is not found in B. In this way we have explained the "common" lines, and consequently we have a purer spectrum B. By comparing B in a similar manner with other metallic spectra we can hope to explain further coincidences and obtain a pure spectrum B.

Suppose, however, that in comparing A and B we reached the stage mentioned above and then noticed that a "short" line of A is found in B. We are not justified in eliminating this line from B—because the lines intermediate between it and the last "longest" line of A common to B are absent from B. Lockyer found many such common lines, which he felt had to be explained and not just ignored.

Since therefore these lines which were common to two or more spectra could not be traced to impurities, what was their probable origin? Their number was so great that to attribute them to physical coincidences, and to rest and be thankful accordingly, would have been to take the very pith and marrow out of the science of spectrum analysis, which we have heard so often is based absolutely upon different substances giving us spectra with special lines for each. The matter then was worthy of serious investigation.[170]

But before considering Lockyer's solution it is necessary to mention that he had also applied the method of long and short lines in studying compound spectra. Each compound has "long" and "short" lines. Knowing the "long" lines of a compound and those of its constituent metallic element, Lockyer could then observe the progress of the dissociation of the compound into its elements. He states on experimental evidence that "the number of true metallic lines which . . . appear is a measure of the quantity of the metal resulting from the dissociation, *and as the metal lines increase in number, the compound bands thin out.*"[171]

Returning now to the "basic" lines, Lockyer succeeded in explaining them by elaborating the dissociation hypothesis. This was in 1879. He now states *that the elements themselves, or at all events some of them, are compound bodies.*"[172] And the lines are called "basic" because they are produced by certain of the "basic molecules representing bases of the so-called elements."[173]

In presenting his ideas Lockyer considers a series of furnaces *A, B, C,* and *D,* of successively lower temperatures (Fig. 4). We are to suppose

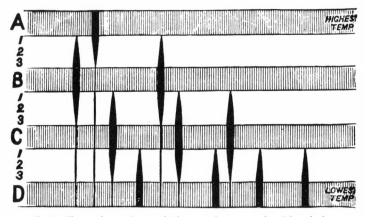

Fig. 4. Illustration of Spectral Changes Accompanying Dissociation

170 *Ibid.,* p. 235. See also: J. N. Lockyer, "On the necessity of a new departure in spectrum analysis," *Nature,* 21 (1879–1880), 5–8; p. 6.
171 J. N. Lockyer, *Chemistry of the Sun,* p. 160.
172 J. N. Lockyer, "Researches in spectrum analysis in connexion with the spectrum of the sun. No. 7. Discussion of the working hypothesis that the so-called elements are compound bodies," *Proc. Roy. Soc.,* 28 (1879): 157–80; p. 159.
173 J. N. Lockyer, "On the necessity for a new departure in spectrum analysis," p. 6.

that in A there is an elementary body a which is capable of forming a compound body β by union with itself "or with something else" when the temperature is lowered. We are further to suppose that the compound β exists alone in furnace B and that its spectrum is the only one visible in B, just as the spectrum of the assumed elementary body a is the only one visible in A. Similarly, in furnaces C and D other compound bodies γ and δ are formed and the same considerations apply.

If some of the compound γ, say, is placed in furnace A, an integrated spectrum composed of the spectra of the three subsances, a, β and γ, will be given. Initially the lines of γ will be "thickest," then those of β, and finally only a will be present and the spectrum will be one of the "utmost simplicity."

Considering next the normal state of any of the other furnaces B, C, and D, Lockyer says that due to partial dissociation lines of the less compounded bodies of the higher temperature furnaces will be present in addition to the lines of the body characteristic of that furnace. For example, we will have lines of β appearing in the spectrum of furnace C, but the lines will be "thin." "In short, the line of a strong in A is *basic* in B, C, and D, the lines in B are basic in C and D, and so on."[174]

Thus the "basic lines" found with metallic spectra are to be attributed to dissociation products of the elements. There is "no spectroscopic break between acknowledged compounds and the supposed elementary bodies."[175] By way of illustration Lockyer says that the differences between calcium itself at different temperatures are as great as when we pass from known compounds of calcium to calcium itself. "There is perfect continuity of phenomena from one end of the scale of temperature to the other."[176]

Lockyer does not attempt to say how many basic "molecules" there might be, or how many "basic lines" might be produced by a particular basic "molecule." His views are still very general. He does, however, speculate upon the nature of the basic "molecules." He desires "to obtain a clear mental view of the manner in which, on *the principle of evolution*, various bases might now be formed, and then become basic themselves."[177] But if he succeeded in finding such a view his expression of it is somewhat obscure—as is not unusual with Lockyer. Fortunately, he will later express his ideas more clearly, so it need only be said here that he suggests three ways of producing "molecules" of

---

174 J. N. Lockyer, "Researches in spectrum analysis in connexion with the spectrum of the sun. No. 7," p. 162.
175 *Ibid.*, p. 165.
176 *Ibid.*
177 *Ibid.*, p. 166.

higher complexity. We have: first, successive additions of the same unit A; second, addition of the two different units A and B and then successive additions of the compound unit to give $(A + B)_2$, $(A + B)_3$, . . . ; and third, successive additions of the unit B to the compound unit $(A + B)$. But this is hardly enlightening—there is no reason why we could not have any one of a number of other schemes. Perhaps the elements could be dissociated, but their composition and structure were still very much of a mystery.

Nevertheless, having given a plausible explanation of the basic lines, Lockyer now sought to confirm it. He soon realized that the sun afforded a means of studying the basic lines at various temperatures.

It was generally accepted that spots and prominences were the hottest regions of the sun, and Lockyer proposed to test the nature of the basic lines by comparing their appearance in these regions with that in the normal solar atmosphere.

Supposing them to represent mere chance coincidences . . . there is no reason why they should vary together when the temperature is changed; while, if they be truly basic, they *must* vary with temperature. Further, they must vary in such a way that other conditions being equal, they shall become stronger when the temperature is increased, and become fainter when the temperature is reduced.[178]

Selecting eighteen solar lines which were common to two or more spectra Lockyer studied their behavior in this way. And he found the "very remarkable fact" that "*these basic lines are always widened in spots.*"[179] The same held true for prominences.

So far as my knowledge of these matters goes, I can imagine no severer test to apply to the hypothesis that the basic lines in the above table are produced by the dissociation of the metals to which the lines are common—in this case chiefly to the metals of the iron group—in the hottest region of the sun, and to my mind the proof is conclusive that at that temperature we have a mixed mass of vapours in which the base is more predominant than the so-called elements.[180]

Lockyer was unable to make further progress with the basic lines, but he was more than satisfied. During the following year (1880) he returned to laboratory considerations of the dissociation hypothesis.[181] In presenting his results he states the hypothesis before giving the obser-

[178] J. N. Lockyer, "On the necessity for a new departure in spectrum analysis," p. 6.
[179] *Ibid.*, p. 7.
[180] *Ibid.*, p. 8.
[181] J. N. Lockyer, "On a new method of spectrum observation," *Proc. Roy. Soc.*, 30 (1880): 22–31.

vations, in such a way that the hypothesis appears to be confirmed. However, all that he can legitimately claim is that the observations may be explained on the hypothesis. I do not say that this is deliberate on Lockyer's part, but the practice is met with occasionally in his work.

In the experiment an electric discharge is passed through a flame in which a metallic compound is vaporized. Lockyer compares the spectrum thus obtained with the spectrum of the flame when the discharge is not passed through. That is, he compares a spectrum of higher temperature with one of lower temperature. Compounds of several metals were employed, but according to Lockyer the magnesium spectra provide the best illustration of the phenomenon. In the magnesium flame spectrum the two least refrangible b lines are seen together with a less refrangible third line. A series of flutings and a blue line are also seen. Upon passing the discharge, all but the two b lines disappear, while an entirely new line makes its appearance. These latter are the only magnesium lines seen in the solar spectrum.

Lockyer's explanation is of course that different "molecules" are involved. The b lines in the flame spectrum are due to "molecules" produced with difficulty at flame temperature, while the other lines are due to "molecules" produced at this or lower temperature. The two b lines are basic, while the others are not, being the remnant of a lower temperature spectrum.

The idea presented by this work—that a spectrum is produced not by one but by several different molecules—was one which was accepted by many of the leading spectroscopists, including Heinrich Kayser and Carl Runge.[182] It had considerable appeal when one considered, say, the complex spectra of iron and cerium—for a spectroscopist of the second half of the nineteenth century found it almost impossible to believe that a single molecule could be so complex as to produce either of these spectra. Thus, George Liveing and James Dewar who, as we shall presently see, were opposed to the dissociation hypothesis, described the iron spectrum as having "a number of vibrations so immense that we can hardly conceive any single molecule to be capable of all of them, and were almost driven to ascribe them to a mixture of differing molecules, though we have as yet no independent evidence of this."[183]

In 1880 Lockyer turned to the molecular interpretation of multiple spectra for support of this idea "that in the case of many bodies the complexity, and therefore the number, of the molecular groupings

[182] H. Kayser and C. Runge, *Ueber die Spectren der Elemente. Erster Abschnitt* (Berlin, 1890), pp. 4–5.
[183] Quoted by J. N. Lockyer in *Chemistry of the Sun,* p. 401.

which give rise to that compound whole called a line spectrum, is considerable."[184] For if line and band spectra of the one element can be attributed to different groupings then "the fact of other molecular groupings giving rise to a complex line spectrum can be more readily accepted, contrary though it is to modern 'chemical philosophy,' as taught at all events in the text books."[185] It is interesting to note that the interpretation of multiple spectra and the dissociation hypothesis were both constructed upon analogy with the behavior of compound spectra. But now, in 1880, the molecular theory of multiple spectra was widely accepted and had almost the appearance of fact.

In the summer of 1881 Lockyer published a comprehensive summary of his dissociation studies in the journal *Nature,* of which he was editor. The idea of an evolution of the elements had been mentioned frequently in earlier papers; but here he introduces what he terms the new theory of chemical evolution, which is said to be but a slight extension of the contemporary chemical view. Chemists, says Lockyer, regard matter as composed of atoms and molecules.

The view now brought forward simply expands the series into a large number of terms, and suggests that the molecular grouping of a chemical substance may be simplified almost without limit if the temperature be increased. A diagram (Figure [5]) will show exactly what I mean. If we assume a very great difference

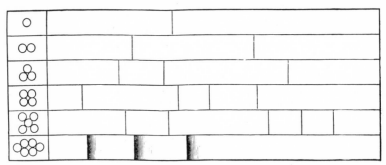

Fig. 5. Diagram Showing How the Evolution of Chemical "Forms" May be Indicated by Their Spectra

in the temperature which can be brought to bear upon a substance we may assume that at the highest temperature we have, for simplicity's sake say, a certain line represented by a single circle; let us imagine the temperature reduced, we shall then get another spectrum, which we can represent by a double circle, if we like to assume that the evolution is one which proceeds by constant additions of the original unit. Coming lower down, we get another substance

---

184 J. N. Lockyer, "On multiple spectra," *Nature,* 22 (1880): 4–7, 309–12, 562–65; p. 4.
185 *Ibid.,* p. 5.

formed with a more complex spectrum represented by three circles; lower down still we have one represented by four circles, another by five, another by six, and so on. We might take another supposition, easier perhaps to some minds, and suppose that evolution proceeded, not by the addition of the initial unit, but by the constant doubling of the substance of the molecule itself. Instead, therefore, of our circles increasing by one, we shall have one, two, four, eight, sixteen, thirty-two, and it will be readily understood that if there are a considerable number of stages of temperature, both within our ken and beyond our ken, and if some substances form themselves perpetually by doubling, then the unit with which we can experiment at low temperature, call it the chemical atom or the chemical molecule, or what you will, must be a very complex thing indeed. If the lower spectrum represents that of a complex body such as iron, or a salt of calcium, the upper spectra will represent those due to the finer groupings brought about by higher temperatures.[186]

As will be seen in a later chapter, the vortex atom was in vogue in British physical science during the last quarter of the nineteenth century, and perhaps this is what Lockyer has in mind when he speaks of "circles." The vortex atom theory, however, was inherently contradictory to the dissociation hypothesis, for vortex atoms were considered indestructible. This no doubt added to the difficulties, to be considered presently, of the hypothesis.

In advancing these views Lockyer was influenced by, in addition to evolutionary thought, yet another prominent branch of nineteenth-century science, organic chemistry. He looked in particular to the hydrocarbon series.

$$
\begin{aligned}
&\text{C} \quad \text{H} \\
&\text{C} + \text{H} \\
&\text{C} + \text{HH} \\
&\text{C} + \text{(HH)(HH)} \\
&\qquad = \dots \text{CH}_4 \\
&+ \text{CH}_2 = \dots \text{C}_2\text{H}_6 \\
&+ \text{CH}_2 = \dots \text{C}_3\text{H}_8 \\
&+ \text{CH}_2 = \dots \text{C}_4\text{H}_{10} \\
&+ \text{CH}_2 = \dots \text{C}_5\text{H}_{12} \\
&+ \text{CH}_2 = \dots \text{C}_6\text{H}_{14} \\
&+ \text{CH}_2 = \dots \text{C}_7\text{H}_{16} \\
&+ \text{CH}_2 = \dots \text{C}_8\text{H}_{18} \\
&+ \text{CH}_2 = \dots \text{C}_9\text{H}_{20} \\
&+ \text{CH}_2 = \dots \text{C}_{10}\text{H}_{22} \\
&+ \text{CH}_2 = \dots \text{C}_{11}\text{H}_{24} \\
&+ \text{CH}_2 = \dots \text{C}_{12}\text{H}_{26} \\
&+ \text{CH}_2 = \dots \text{C}_{13}\text{H}_{28} \\
&+ \text{CH}_2 = \dots \text{C}_{14}\text{H}_{30} \\
&+ \text{CH}_2 = \dots \text{C}_{15}\text{H}_{32} \\
&+ \text{CH}_2 = \dots \text{C}_{16}\text{H}_{34}
\end{aligned}
$$

[186] J. N. Lockyer, "Solar physics: chemistry of the sun," *Nature,* 24 (1881): 267–74, 296–301, 315–24, 365–70, 391–99; pp. 394–95.

And by direct analogy Lockyer proposed an evolution of the elements as follows:

$$
\begin{array}{ll}
a \quad b & \\
a + b & \\
a + bb & \\
a + (bb)(bb) & \\
\quad = \ldots ab_4 & \\
+ ab_2 = \ldots\ldots a_2b_6 & \\
+ ab_2 = \ldots\ldots a_3b_8 & \\
+ ab_2 = \ldots\ldots a_4b_{10} & \\
+ ab_2 = \ldots\ldots a_5b_{12} & \\
+ ab_2 = \ldots\ldots a_6b_{14} & \\
+ ab_2 = \ldots\ldots a_7b_{16} & \\
+ ab_2 = \ldots\ldots a_8b_{18} & \\
+ ab_2 = \ldots\ldots a_9b_{20} & \\
+ ab_2 = \ldots\ldots a_{10}b_{22} & \\
+ ab_2 = \ldots\ldots a_{11}b_{24} & \\
+ ab_2 = \ldots\ldots a_{12}b_{26} & \\
+ ab_2 = \ldots\ldots a_{13}b_{28} & \\
+ ab_2 = \ldots\ldots a_{14}b_{30} & \\
+ ab_2 = \ldots\ldots a_{15}b_{32} & \\
+ ab_2 = \ldots\ldots a_{16}b_{34} & \\
\end{array}
$$

This diagram deals with a very simple case of evolution, and it deals with this evolutionary process, going along a single line. Of course we know very well that in the organic kingdom evolution always proceeds along many lines, but to simplify the problem I have dealt with one of the simplest that I can think of. Let us assume that in a certain hottest star there shall be two substances, which we will call a and b. They will first at the transcendental temperature which I assume, exist as separate entities; the temperature being then reduced, they probably will combine, and, instead of two atoms, a and b, we shall have one group of $a + b$. If the temperature is still further reduced, we shall get b combining with b; in that case we shall have a grouping consisting of $a + 2b$. Let the same operation be performed again, we shall then have $a + 4b$, combining into two groups of 2; we shall have what we can represent, in short, in chemical language ab_4. Now, having got out ab_2, having got our temperature reduced, let us assume that ab_2 is now the substance linked on to give a greater complexity, instead of b or $2b$ merely. We then have this series given in the table.[187]

Lockyer says that:

taking this simple case we are justified in saying that if nature, in the regions which we cannot get at, works in the same way as she does in the regions which we can get at, the view is not absurd, and in fact any one who wishes to dispute the view in such a case as this has, I think, the *onus probandi* thrown upon him. He must show that either in a certain latitude or longitude, or at a certain temperature, or under some unknown condition the laws of nature are absolutely changed, and give place to new ones. That has not yet been found in any other region of natural philosophy.[188]

187 *Ibid.*, p. 395.
188 *Ibid.*

Thus Lockyer was not saying that the elements are built up as the hydrocarbons, but that starting with simple evolutionary ideas and the hydrocarbon series, and accepting his idea of the uniformity of nature, a not impossible view of the composition of the elements could be suggested.

It is surprising that Lockyer apparently made no effort to make his scheme comply with the periodic table of the elements. Nevertheless, it was an attempt to render the general physical view of matter more concrete. He next mentions or quotes "chemical philosophers" who had previously also advocated this view.

First, Dalton is quoted: "We do not know that any of the bodies denominated elementary is absolutely indecomposable."[189] Thomas Graham follows: "It is conceivable that the various kinds of matter now recognized in different elementary substances may possess one and the same element or atomic molecule existing in different conditions of movability."[190] Then Lockyer mentions Hermann Kopp, J. B. Dumas, and Sir Benjamin Brodie, formerly Professor of Chemistry at Oxford. Lockyer calls Dumas the "greatest chemical philosopher now living," which perhaps is not surprising in view of the fact that from the 1830s Dumas had been a supporter of the radical theory of the elements. This latter was constructed on analogy with the homologous series of organic chemistry.

Lockyer continues:

If, then, the three greatest English chemists we can name, and the most eminent philosophers in France and Germany, give their opinion in behalf of the compound nature of the chemical elements, can these simpler forms be any other than those we detect by means of the spectroscope? By the conditions of the problem and the absence of knowledge they are not decomposable in the laboratory; if they were they would cease to be elementary bodies at once, and would be wiped out of our tables. Nor do I think it possible that in the present state of our knowledge they can be revealed to us in any other way than by the spectroscope.[191]

Lockyer also adds that Maxwell allowed for the possibility of elements becoming decomposed at the high temperatures of the sun and stars. He could also have mentioned Lothar Meyer, Michael Faraday, Stokes, and Herbert Spencer as supporters of the idea of compound elements. The last three were quoted in 1885 by William Crookes, when, in his presidential address to the Chemical Section of the British Association for the Advancement of Science, he too argued for an

189 *Ibid.*, pp. 395–96.
190 *Ibid.*
191 *Ibid.*, p. 396.

evolution of the elements. According to Crookes the notion of the "complexity of our supposed elements" was "so to speak, in the air of science."[192]

Thus we see that in arguing for the compound nature of the elements, Lockyer was not proposing anything really new. What was new though, were his apparent demonstrations of the dissociation of the elements and consequently of the correctness of this view.

Thus far I have described the general development of Lockyer's dissociation hypothesis over a period of some seven years and have not yet mentioned any contemporary reactions to it. I have done so because by 1881 Lockyer's ideas were fully developed and little of importance was added during the period down to 1897. Meanwhile, however, considerable interest had been generated in the dissociation hypothesis. For example, prior to 1878 the Dutch Society of Science had encouraged its further investigation by offering a prize.[193]

Furthermore, the dissociation hypothesis was not without a rival. Most of the phenomena which it purported to explain could also be satisfactorily explained by the theory of molecular collisions, or bell hypothesis,[194] met with in the previous section in connection with Schuster's discussion of the oxygen spectra. This theory had existed before the dissociation hypothesis was proposed. It is not clear that it originated with any one individual, but it may be found in Ångstrom's work where he endeavors to explain the appearance of new lines when the temperature of an incandescent gas is raised. The idea is that the lines are produced by the more violent collisions occurring between molecules at the higher temperature.

Lockyer was well aware of the alternative hypothesis and repeatedly rejected it. He thought it quite plausible in explaining slight differences between spectra, but then the differences he was concerned with were, in his opinion, not slight. For example, there was the "extraordinary fact" that the flame and solar-spot spectra of iron had no lines in common. Furthermore, as Lockyer viewed the matter, the law of continuity was on his side. For the change in spectra in going from what he considered to be one form of calcium, say, to another (disso-

192 W. Crookes, *Rep. British Assn.* (1886): 558–76; p. 560. In this connection see also the recent article by W. V. Farrar, "Nineteenth-century speculations on the complexity of the chemical elements," *British J. Hist. Sc.*, 2 (1965): 297–323, in which Lockyer is only briefly mentioned.

193 Mentioned by Lockyer in *SSA*, p. 197.

194 "Bell hypothesis" because, as Lockyer explains, "many high notes, both true and false, . . . can be produced out of a bell without its fundamental one." J. N. Lockyer, "Researches in spectrum analysis in connexion with the spectrum of the sun. No. 7," p. 165.

ciated) form is "the same in kind and about the same in degree" as the change in spectra in passing from a calcium compound to calcium.[195]

Lockyer found himself faced with a more serious difficulty when in 1880 that part of his work dealing with the basic lines of the solar spectrum first came under attack. Using the best grating he had yet seen—one by Rutherfurd—Charles Augustus Young, Professor of Astronomy at Princeton, had examined the seventy solar lines given by Ångstrom as common to two or more elemental spectra. Of these, fifty-six were now seen to be either double or triple, seven appeared to be single, while the nature of the remaining seven was uncertain. Thus it appeared that most of the coincident lines called basic lines by Lockyer might really be only approximately coincident. And in order to decide the matter Young urged that the bright line spectra concerned be compared under "enormous" dispersion. "If, in this research it should be found that *both* of the components of a double line were represented in the spectra of two different metals, and the suspicion of impurity were excluded, we should then indeed have a most powerful argument in favour of some identity of material or architecture in the molecules of the two substances involved."[196]

In replying to Young, Lockyer developed a very obscure probability argument on the distribution of double solar lines, which need not be reproduced. We need merely note his conclusion that "the idea that these basic lines are simple creatures of the imagination, simply chance coincidences, will not really stand at all."[197]

However, two Cambridge men, George Liveing and James Dewar, were not so sure, and undertook the task suggested by Young of comparing the metallic spectra. Liveing was professor of chemistry and Dewar, Jacksonian professor of experimental philosophy, and together they arrived at many valuable experimental spectroscopic results during the last quarter of the nineteenth century. Having examined all but a few of the "supposed" coincidences they reported in 1881 that:

The results which we have recorded strongly confirm Young's observations, and leave, we think, little doubt that the few as yet unresolved coincidences either will yield to a higher dispersion, or are merely accidental. It would indeed be strange, if amongst all the variety of chemical elements, and the still greater variety of vibrations which some of them are capable of taking up, there were no two which could take up vibrations of the same period. We certainly should

---

195 J. N. Lockyer, "Solar physics: chemistry of the sun," p. 394. See also J. N. Lockyer, "Researches in spectrum analysis in connexion with the spectrum of the sun. No. 7," p. 165.
196 C. A. Young, "Spectroscopic notes, 1879–1880," *American J. Sc.*, 20 (1880): 353–58; p. 356.
197 J. N. Lockyer, "Solar physics: chemistry of the sun," p. 393.

have supposed that substances like iron and titanium, with such a large number of lines must each consist of more than one kind of molecule, and that not single lines, but several lines of each, would be found repeated in the spectrum of some other chemical elements. The fact that hardly single coincidences can be established is a strong argument that the materials of iron and titanium, even if they be not homogeneous, are still different from those of other chemical elements. The supposition that the different elements may be resolved into simpler constituents, or into a single one, has long been a favourite speculation with chemists; but however probable this hypothesis may appear, *a priori,* it must be acknowledged that the facts derived from the most powerful method of analytical investigation yet devised give it scant support.[198]

Again we meet with the idea that the more complex spectra may be produced by more than one molecule. The nature of these molecules is not specified, but elsewhere Liveing and Dewar speak of the oxygen molecule O_4, in addition to O_3, and O_2.[199] So presumably we may have a series of iron molecules Fe, Fe_2, Fe_3, Fe_4, . . . producing the iron line spectrum. Complexity of spectra is to be explained in terms of several molecules of higher complexity. Clearly spectroscopists saw a limit to the complexity of the spectral emitters, and we shall have an opportunity of corroborating this shortly. I need only add here that I have not found any physicist-spectroscopists attempting to describe the differences between molecules producing band spectra and those producing line spectra.

Most people seem to have accepted the result that the basic lines were spurious. But the fact that they were did not invalidate the dissociation hypothesis which had been used to explain them. Nevertheless, as Lockyer had said, the behavior of the basic lines in sunspots had afforded the severest test for the hypothesis, and consequently it now lost much of its appeal.

But there were still laboratory phenomena which could be explained by, or alternatively lend support to, the dissociation hypothesis. However, Liveing and Dewar also regarded the hypothesis "as being without any foundation in fact."[200] And in the same year (1881) they confounded what Lockyer had taken to be the best laboratory illustration of dissociation—the behavior of the magnesium spectrum.

Liveing and Dewar showed that Lockyer had worked not with the

198 G. D. Liveing and J. Dewar, "On the identity of spectral lines of different elements," *Proc. Roy. Soc.,* 32 (1881): 225–30; in G. D. Liveing and Sir J. Dewar, *Collected Papers on Spectroscopy* (Cambridge, 1915), pp. 133–39; pp. 138–39.

199 G. D. Liveing and J. Dewar, "Notes on the absorption-spectra of oxygen and some of its compounds," *Proc. Roy. Soc.,* 46 (1889): 222–30; in *Collected Papers . . .,* pp. 382–90; p. 387.

200 G. D. Liveing and J. Dewar, "On the spectra of magnesium and lithium," *Proc. Roy. Soc.,* 30 (1880): 93–99; in *Collected Papers . . .,* pp. 78–84; p. 79.

spectrum of magnesium, as he had thought, but with that of a magnesium-hydrogen compound.[201] Furthermore, they compared the true flame, arc, and spark spectra of magnesium and found that in addition to their having common lines all three were quite similar. This is contrary to what ought to be the case according to Lockyer.

Liveing and Dewar accepted the collision theory as one possibility—"the chemical atoms of magnesium are either themselves capable of taking up a great variety of vibrations, or are capable by mutual action on each other, or on particles of matter of other kind, of giving rise to a great variety of vibrations of the luminiferous aether."[202] They were never again to concern themselves with Lockyer's dissociation hypothesis.

Yet another criticism was made in 1883 by Walter Noel Hartley, Professor of Chemistry at the Royal College of Science, Dublin, who understood Lockyer to say that every elementary substance might be decomposed into as many simple substances as there are lines in its spectrum.[203] No doubt this refers to Lockyer's diagram illustrating his proposed theory of evolution of the elements. Hartley objects that "with even very moderate dispersive power something like 1,200 lines can be recognized in the spectrum of iron, an element which has an atomic weight of 56, and it is simply inconceivable that a body of the chemical nature of iron can have a molecular structure so complex as to be composed of 1,200 different simpler substances." Thus while the elements might be compound, the number of components had to be relatively small. Hartley did not suggest any way out of the difficulty, unlike Liveing and Dewar. But all three were agreed that the iron molecule could not be as complex as its spectrum suggested. In contrast, Heinrich Kayser, who in 1883 thought the dissociation hypothesis an open question, could assume that the iron molecule consisted of 5,000 atoms, whether of the one or different types, and further that it could be dissociated into two molecules—one consisting of 3,000 and the other of 2,000 atoms.[204]

In 1887 Lockyer brought together, in his book *Chemistry of the Sun,* all of his previously published material on the dissociation hypothesis. One eagerly looks in this for Lockyer's replies to his critics, but one finds the material presented almost exactly as before with never a

201 G. D. Liveing and J. Dewar, "Investigations on the spectrum of magnesium, No. 1," *Proc. Roy. Soc., 32* (1881): 189–203; in *Collected Papers . . .,* pp. 118–33.
202 *Ibid.,* p. 182.
203 W. N. Hartley, "On homologous spectra," *J. Chem. Soc. London, 43* (1883): 390–400; p. 398.
204 H. Kayser, *Lehrbuch der Spektralanalyse* (Berlin, 1883), pp. 212, 210.

mention of the criticisms. Indeed it was not until twelve years later that Lockyer admitted his work on the basic lines of the solar spectrum to be "universally rejected."[205] According to Heinrich Kayser, who continued to keep an open mind on the dissociation question, it was because Lockyer ignored these criticisms and continued as if the nonexistence of the basic lines had not been demonstrated that his theory became "greatly damaged" and "discredited."[206]

The existence of the basic lines was again denied in 1890, this time by Kayser and Carl Runge, who armed with a Rowland grating were in the process of composing the most accurate metallic spectral maps yet produced.[207] Their rival in the spectral series field, Janne Rydberg, had also rejected the basic lines in the previous year. In addition Rydberg considered the "second support" of the dissociation hypothesis also to have been vitiated. For he had apparently found that the new lines appearing in the sodium and potassium spectra at higher temperatures "belong as harmonics, either to the same series of vibrations or to series which are inseparable from one another."[208]

But the dissociation hypothesis continued to have its occasional supporters in addition to Lockyer.[209] And one of the reasons for its continued support must be that dissociation was implicit in the physical view of matter. As Schuster remarked in 1897: "Most of us are convinced in our innermost hearts that matter is ultimately of one kind, whatever ideas we may have formed as to the nature of the primordial substance."[210] Schuster, however, was not one to be confused and demanded a "direct" demonstration of the dissociation hypothesis. "If Mr. Lockyer is right we must look forward to finding some trace of helium, or calcium, or hydrogen in the discharge taken from iron poles. When that is done, and not till then, will his theory be considered as proved."[211]

It must be added that Lockyer was not the only spectroscopist to

---

205 J. N. Lockyer, "On the chemistry of the hottest stars," *Proc. Roy. Soc.*, 61 (1897): 148–209; p. 198.

206 H. Kayser, *Handbuch der Spectroscopie*, 2: 266.

207 H. Kayser and C. Runge, *Über die Spectren der Elemente. Dritter Abschnitt* (Berlin, 1890), p. 16.

208 J. R. Rydberg, "Recherches sur la constitution des spectres d'émission des éléments chimiques," *Konliga Svenska Vetenskaps-Akademiens Handligar*, 23 1888–89): no. 11, p. 11.

209 See, for example, W. C. Roberts-Austen, "Metals at high temperatures," *Proc. Royal Institution*, 13 (1892): 502–18; p. 509; and, J. M. Eder and E. Valenta, "Ueber die verschiedenen Spectren des Quecksilbers," *Denkschriften der Kaiserlichen Akad. der Wissen. Wien.*, 61 (1894): 401–30; p. 429.

210 A. Schuster, "On the chemical constitutions of stars," *Proc. Roy. Soc.*, 61 (1897): 209–13; p. 212.

211 *Ibid.*, p. 213.

claim to have demonstrated, to his own satisfaction at least, the theory that the elements are of a compound nature. There were at least two others during the last quarter of the century. However, Lockyer's position was the most general—in regard to the specific nature of the components of elements—and the best founded in fact, and for these reasons the most plausible and enduring. We shall see in the following chapter that in 1880 G. L. Ciamician considered the elements of the periodic table to be compounds of one of hydrogen, nitrogen, fluorine and carbon, together with different amounts of oxygen in various states of condensation. Significantly, however, he did not claim these five elements to be the "last components of matter." Ciamician's teaching was short-lived as it proved to be based on highly inaccurate determinations of spectra.

The same can be said of Anton Grünwald's similar speculations. Grünwald was a mathematics professor at the University of Prague and his first two papers on "mathematical spectrum analysis" appeared in 1887 and 1888.[212] Immediately thereafter his work was severely criticized by Joseph Sweetman Ames of The Johns Hopkins University and particularly by Kayser, who later was to describe Grünwald's papers as "perhaps the most unscientific ever made in a physical field."[213] From a comparison of spectra Grünwald claimed that all of the "so-called" elements were compounds of two primary gaseous elements which he denoted by a and b. Hydrogen, for example, was a compound of one volume of b and four of a.

[212] A. K. Grünwald, "Ueber die merkwürdigen Beziehungen zwischen dem Spektrum des Wasserdampfes und den Linienspectren des Wasserstoffs und Sauerstoffs, sowie ueber die chemische Structur der beiden letzten und ihre Dissociation in der Sonnenatmosphäre," *Astronomische Nachrichten,* 117 (1887): 201–14; in *Chem. News,* 56 (1887): 186–88, 201–2, 223–24, 232. Also: A. K. Grünwald, "Mathematische Spectralanalyse des Magnesiums und der Kohle," *Berichte Akad. Wissen. Wien,* 96 (1888): 1154–1216; in *Phil. Mag.,* 25 (1888): 343–50.

[213] H. Kayser, *Handbuch der Spectroscopie,* 2: 588. Also: J. S. Ames, "Grünwald's mathematical spectrum analysis," *Nature,* 40 (1889): 19; and H. Kayser, "Ueber Grünwald's mathematische Spectralanalyse," *Chemiker Zeitung,* 13 (1889): 1655 and especially 1687–89.

III

||||||||||||||||||||||||||||||||

Quest for Spectral Series Formulae

THE GRADUAL GROWTH of physical interest in spectra in relation to atoms and molecules had meanwhile also been taking place in a second direction. Physicists had taken up the task, commenced by chemists, of seeking numerical relationships between different spectra and between the different lines of individual spectra. It was hoped that some knowledge of atoms and molecules might in this way be acquired.

This movement began about 1869, but as early as 1864 at least one investigator had recognized numerical relationships between the lines of certain spectra. This, however, appears to be an isolated case and not a part of that continuous thread of investigation with which this chapter is concerned. Nevertheless, it is not unconnected with later developments and will be briefly considered.

In chapter II we saw that in 1862 Alexander Mitscherlich discovered that chemical compounds, like metals and gases, have characteristic spectra. During his further researches in this area Mitscherlich noticed that the halogen compounds of the alkaline earths, with the exception of the fluorides, gave spectra which displayed corresponding single lines.[1] These compounds of calcium, strontium, and barium were the only ones which he found to give line spectra as opposed to band spectra.

With the barium compounds Mitscherlich observed that the spacings of the two principal lines found in the spectra of the chloride, iodide, and bromide, were to each other as the atomic weights of these compounds. Thus knowing the spacing of the two principal lines in the chloride spectrum to be 3.9 degrees, measured on his arbitrary scale,

1 A. Mitscherlich, "On the spectra of compounds of simple substances," *Phil. Mag.*, 28 (1864): 169–89; p. 182.

he could calculate for example the spacing, x, of the two principal lines in the barium iodide spectrum from the equation:

$$\frac{3.9}{x} = \frac{\text{Atomic weight of barium chloride}}{\text{Atomic weight of barium iodide}}$$

The other, more feeble, lines of the three barium haloids could be related in a similar manner. On the other hand, the spacing of corresponding lines in the spectra of the calcium haloids varied inversely as the atomic weight. And the same was found to hold for the strontium haloids.

However, while the spectra of metallic oxides and the spectra of various compounds of a given metal—as, for example, the copper haloids—displayed certain similarities, Mitscherlich could find no numerical relationships analogous to those found with the spectra of the alkaline earth haloids. Nevertheless, he was confident that such would be found to exist among the spectra of the metallic oxides.

Only in his concluding paragraph does Mitscherlich refer to the spectra of metals, as opposed to the spectra of their compounds. He has not "been able to find properties of the spectra of metals which enable a connection of the metals *with one another* to be recognized."[2] But again, such a connection seems to be indicated, for zinc and cadmium, which have similar chemical properties, have also similar spectra.

Mitscherlich did not pursue these researches further, and at the time no one appears to have paid them much attention. Coming at the end of his major paper on the spectra of chemical compounds they were doubtless eclipsed by the great interest in these latter. Five years later however, in 1869, the subject was taken up again by Éleuthère Mascart, then a physics teacher in the high-school at Versailles. An important problem, he said, "which spectrum analysis ought to consider, is to know if there exists a relation between the different lines of a given substance or else between the spectra of analogous substancs."[3] By 1863 it had been recognized that the sodium spectrum displayed not one (as Bunsen and Kirchhoff had stated three years earlier) but six principal lines, all of which Mascart now observed to be double.[4] Moreover, the spacing of the lines of each doublet was almost equal to that of the D lines.[5] In addition the investigation of ultraviolet spectra, which in the 1860s had begun to be carried out satisfactorily with Mascart as one of

2 *Ibid.*, p. 187.
3 E. Mascart, "Sur les spectres ultra-violet," *Comptes Rendus*, 69 (1869): 337–38.
4 The six principal lines of sodium had been observed by E. Wolf and E. Diacon, "Note sur les spectres des métaux alcalins," *Comptes Rendus*, 55 (1862): 334–36.
5 The terms "doublet" and "triplet" etc. did not come into use until about 1880, but it is convenient to use them here.

the leaders in the field, had revealed similar phenomena. Thus in addition to a triplet in the green, the magnesium spectrum was found to possess a further two triplets in the ultraviolet, each of which was strikingly similar to that of the first triplet.

With these and other instances in mind Mascart remarks that it seems difficult to believe "that the reproduction of such a phenomenon [doublet or triplet] is a matter of chance; is it not more natural to admit that these groups of similar lines are harmonics which relate to the molecular constitution of the luminous gas?"[6] Yet he was well aware that a great number of similar observations would be necessary if one was to discover the laws governing these "harmonics." The latter term is not explained, but it is clear that Mascart has in mind the widely adopted analogy with sound.[7]

Mascart's paper may be regarded as marking the beginning of a continuing interest which led to the discovery of the first spectral series formula some fifteen years later. In considering, in the present chapter, the history of this interest down to 1897, I have again found it convenient to divide the work into several sections.

1. Spectral Relationship and the Periodic Table

Commencing about the time of Mascart's paper, two lines of spectral research may be distinguished, one pursued by chemists and the other by physicists. Each was largely independent of the other and continued to be so until united some twenty years later in the work of J. R. Rydberg. The chemists' approach contrasts sharply with that of the physicists, as will be fully appreciated when the latter is considered in the next section. Here the nature and results of the chemists' researches will be sketched in general rather than closely considered. My reason for doing so is that for the most part this work was later to be severely criticized by the more mathematically-minded physicists as being inexact and largely incorrect. Because of its semi-quantitative nature it was of no direct importance to the computation of spectral series formulae. Nevertheless, the work forms an integral part of our story and was certainly closely examined by those physicists whose work will subsequently be considered. In addition it reveals an interesting

---

[6] E. Mascart, "Sur les spectres ultra-violet," p. 338.

[7] In addition to those authors mentioned in chapter I, Bunsen also wrote of an analogy with sound. Of certain spectral absorption phenomena he says: "They remind me of the slight and gradual alterations in pitch which the notes from a vibrating elastic rod undergo when the rod is weighted, or the change of tone which an organ-pipe exhibits when the tube is lengthened." "On the phenomena observed in the absorption-spectrum of didyium," *Phil. Mag.*, 32 (1866): 177–82; p. 182.

interaction of two developing branches of nineteenth-century science. The principal effect of Mascart's paper was to elicit immediately from Francois Lecoq de Boisbaudran, a chemist of private means, a series of papers which were considered important enough for the Académie des Sciences to take the unusual step of publishing in full in the *Comptes Rendus.* This is surprising, for while the papers contain several important experimental observations they also contain much unrestrained speculation.

Apparently the questions posed by Mascart had occurred to Lecoq much earlier, at least before 1865, but at that time spectra in addition to being incompletely known were also only crudely quantitative—as was seen above with Mitscherlich. Indeed, progress in the search for numerical relationships between spectral lines only became possible when spectra were known in much greater detail, and that detail made accurately quantitative. Nevertheless, in 1865 Lecoq had deposited with the Académie des Sciences a sealed letter in which, after having treated other questions, he recorded observations made on the alkali spectra as drawn by Kirchhoff and Bunsen. Now, four years later, he discloses several of these observations, declaring at the outset that he believes he has "the general law which presides over the formation of spectra and explains the singular repetition of the same groups of lines at different degrees of the luminous scale."[8]

Having been "strongly struck" by the appearance of Bunsen and Kirchhoff's alkali spectra, Lecoq had begun to consider "the so intimate relationships which link the properties of bodies to their atomic weight." This was at a time when many chemists were examining these relationships, although not in association with spectra. After more than two years contemplation of the alkali spectra Lecoq had discovered a "new and remarkable relationship":

The spectral lines of the alkali metals (and alkaline earths) classified by their refrangibilities, are placed, as the chemical properties, following the order of atomic weight.[9]

For example, the rubidium spectrum appears analogous to that of potassium, only it is situated more toward the red. Bunsen and Kirchhoff had not given the wavelengths of any lines, but Lecoq believed that if such were known then very interesting numerical comparisons could doubtless be made.

It was later possible for Lecoq, in 1869, to make approximate numer-

8 F. Lecoq de Boisbaudran, "Sur la constitution des spectres lumineux," *Comptes Rendus,* 69 (1869): 445–51; pp. 445–46.
9 *Ibid.,* p. 446.

ical investigations of the similarities of certain spectra and also of the relationships between the lines of a single spectrum. He gives the wavelengths of the four principal lines of potassium and observes that the spacing of consecutive lines are in geometric proportion—11.9: 16.1:23.2.[10] But the physicists would soon reject such crude numerical approximations, and also direct their attention to the wavelengths of the lines rather than to their spacings. Passing to the rubidium spectrum Lecoq finds that the spacings of its four principal lines are also in geometric proportion. If now the potassium and rubidium spectra are compared it is found that each consists of: a double red line; five lines situated toward the middle of the spectrum; and a line, or double line, in the violet. And the wavelength of a rubidium line can be obtained by multiplying the wavelength of the corresponding potassium line by a coefficient which is approximately constant, but which increases slightly for the smaller wavelengths. Analogous relationships are found between the spectra of rubidium and caesium, but the latter's spectrum is too complex for Lecoq to consider in detail.

A similar investigation of the spectra of calcium, strontium, and barium, revealed again a "family likeness." Thus in the same year in which Mendelejeff published his first periodic table of the elements, and the year before Lothar Meyer published his atomic-volumes curve, Lecoq had recognized that in addition to the chemical similarities within certain groups of elements there were also physical similarities.

In investigating the numerical relationships between the individual lines of a given spectrum Lecoq turns to the nitrogen spectrum, which from the studies of Plücker and Hittorf he knows contains two series of bands.[11] And what he does is not to look for relationships within each series, but rather for relationships between the two series. In contrast to this, physicists would later seek and find relationships within, for example, each of the three series of lines in the spectrum of an alkali metal. However, like the physicists, Lecoq this time investigates the relationships between the wavelengths of the bands as opposed to their spacings.

One series is composed of bands, each consisting of two diffuse lines. The second is composed of shaded bands, each consisting of a nebulous luminosity terminated by a diffuse line on the less refrangible side and falling off rapidly in intensity toward the blue. Plücker and Hittorf had observed as many as thirty-five lines in several of the latter bands.

---

[10] F. Lecoq de Boisbaudran, "Sur la constitution des spectres lumineux," *Comptes Rendus,* 69 (1869): 606–15; p. 610.

[11] F. Lecoq de Boisbaudran, "Sur la constitution des spectres lumineux," *Comptes Rendus,* 69 (1869), 694–700. For the work of Plücker and Hittorf see p. 55.

In determining the wavelengths of the bands Lecoq takes the mean of the double lines in the first series, and the center of the principal terminating line in the second series.

Lecoq gives two columns of wavelengths, without distinguishing one series from the other, and says that when the relationships between the wavelengths are calculated it is found that the ratio 3:4 occurs frequently. This leads him to declare that the nitrogen spectrum is probably composed of two harmonics, one represented by double lines and the other by shaded bands. This probability gains strength in Lecoq's eyes from the fact that Plücker and Hittorf have found the two spectra to occur separately—that of double lines at lower, and that of shaded bands at higher, temperatures. And this he says is what ought to happen if the two partial spectra are harmonics, noting further that Plücker and Hittorf have attributed the two series to different allotropes of nitrogen. Finally, an irregularity in the spacing of two bands in one series is reproduced in the spacing of the two corresponding bands of the other. Thus to his own satisfaction Lecoq had established a remarkable fact—what he calls the harmonic relationship of two series.

It is clear that Lecoq's approach to spectra was influenced by the widespread contemporary interest among chemists in the periodic grouping of the elements. In this he was not alone, for within the decade of the 1870s four other chemists pursued a similar line of qualitative research.

Three other French chemists examined the spectra of remaining groups of elements and arrived at a conclusion similar to Lecoq's for the alkali metals and alkaline earths. In 1871 Louis Joseph Troost, a lecturer at the École Normale Supérieure in Paris, and Paul Gabriel Hautefeuille, a teacher of industrial chemistry at the École Centrale des Arts et Manufactures, collaborated in a study of the spectra of the group of carbon, boron, silicon, titanium, and zirconium. They concluded that a comparative study of the spectra of these elements led to a classification identical to that arrived at from a comparison of their other properties.[12] A similar conclusion was reached simultaneously by A. Ditte, also of the École Normale, in his study of the group of oxygen, sulphur, selenium, and tellurium;[13] and again a little later in his study of the groups of: nitrogen, phosphorus, arsenic, antimony, and tin; and chlorine, iodine, and bromine.[14] It should be

[12] L. Troost and P. Hautefeuille, "Sur les spectres du carbone, du bore, du silicium, du titane et du zirconium," *Comptes Rendus,* 73 (1871): 620–22; p. 622.

[13] A. Ditte, "Sur les spectres du soufre, du sélénium et du tellure," *Comptes Rendus,* 73 (1871): 622–24.

[14] A. Ditte, "Sur les spectres des corps appartenant aux familles de l'azote et du chlore," *Comptes Rendus,* 73 (1871): 738–42; p. 742.

noted that neither in Mendelejeff's, Lothar Meyer's, nor the modern periodic table is boron classified with the carbon group or tin with the nitrogen group. But apparently they were in the classification of J. B. Dumas, which all three authors were following—no doubt diplomatically, as Dumas was an influential member of the Académie.

Finally, toward the end of the decade a similar approach was taken by Giacomo Luigi Ciamician, a young Italian studying chemistry at the University of Vienna. Initially Ciamician examined the spectra of thirty-one elements and arrived independently at the same conclusion as Lecoq regarding the similarity of the spectra of a given chemical group. His chemical orientation toward spectra is further seen in his italicized observation that *"the increase of wavelength of homologous lines of related elements accords with a greater intensity of the chemical kinetic energy of the elements."*[15] Thus the most chemically reactive group of elements, the alkalis, have also the lines of greatest wavelength. Ciamician even goes so far as to mention that for a considerable time there have been attempts to describe the properties of elements as functions of their atomic weights, and now it appears that they are also functions of the wavelengths of their spectral lines.

Like Mitscherlich, Ciamician also noted relationships existing between the spectra of the haloids of each of the alkaline earths and between the spectra of other chemical compounds, but unlike Mitscherlich he did not here attempt a quantitative treatment. However, in considering the zinc and copper groups of elements Ciamician shows that the differences of wavelength between corresponding lines of the spectra of elements of the zinc group are multiples of twenty-five, and of the copper group multiples of thirty. It is true that he neither believes these to be the correct relationships, nor even that the laws, which he does believe to exist, will be found to be as simple. But they demonstrate to him that spectroscopy is more than a means of chemical analysis, that it can yield "information on the species of movement of the atom."[16]

Three years later, in 1880, Ciamician considered the similarities of the spectra of chemically related elements more closely. And this time he used the results to speculate on the constitution of the elements which he, as many others, regarded from the standpoint of the so-called radical theory of the elements. Thus, he took the atomic weights of

15 G. L. Ciamician, "Ueber die Spectren der chemischen Elemente und ihrer Verbindungen," *Sitzungsberichte der Mathematisch-Naturwissenschaftlichen Classe der kaiserlichen Akademie der Wissenschaften, Wien*, 76 (1877), Abt., 2, pp. 499–517; p. 514.
16 *Ibid.*, p. 517.

the elements of any group to be the sums of the atomic weight of the first member of the group and a multiple of sixteen (the atomic weight of oxygen). For example, the first halogen, fluorine, has an atomic weight of 19, and the atomic weights of the others are: chlorine—19 + 16; bromine—19 + (4 × 16) ; and iodine—19 + (7 × 16) . Ciamician observed that the spectrum of an element corresponds in part to the spectrum of the corresponding member of the oxygen group, as well as in part to the spectrum of the first element of its group—that is, he saw horizontal as well as vertical relationships among the elements of Mendelejeff's periodic classification. Whence, he concluded, to take the halogens again as example, that fluorine, chlorine, bromine, and iodine consist of varying amounts of the same matter in different degrees of condensation. In this case the constituents are fluorine and oxygen. Similarly, all of the other elements are combinations of hydrogen, carbon, or nitrogen with more or less condensed amounts of oxygen. However, he is not prepared to say that these five basic elements are the "last components of matter."[17]

A decade later J. R. Rydberg in giving an accurate assessment of Ciamician's work revealed the attitude of physicists to it.

As for the comparative studies of Ciamician on luminous spectra, it is to be regretted that he has not attached more importance to the purely experimental part and that he has not attempted to give a more proper form to the rich materials which he has assembled by reducing them to wavelengths. For the analogies which he believes to have found between the spectra of the elements of the same natural group are only chimeras, with the exception of some which had already been recognized by Lecoq or Mitscherlich. There are without doubt analogies between different spectra as far as in the smallest details, but they are not of the nature which Ciamician supposes.[18]

Beginning in the early eighties numerical spectral relationships were to be the concern primarily of physicists, to whose work we now turn.

2. Stoney's Theory of Spectra

The search on the part of physicists for exact numerical relationships between the lines of a given spectrum began in the early 1870s and for several years was conducted mainly by George Johnstone Stoney who, as mentioned in chapter II, was one of those who developed the original molecular theory of spectra.

[17] G. L. Ciamician, "Spectroskopische Untersuchungen," *Sitzungsberichte . . . Wien*, 82 (1880), Abt., 2, pp. 425–27.
[18] J. R. Rydberg, "Recherches sur la constitution des spectres d'émission des éléments chimiques," *Konliga Svenska Vetenskaps-Akademiens Handligar*, 23 (1888–89), No. 11, p. 6.

The fundamental problem to be considered was the structure, often complex, of line spectra produced at ordinary temperatures. If each line of a spectrum corresponds to a specific movement of the emitting molecule how can an iron molecule, say, possibly vibrate in so many different ways? In 1871 Stoney advanced an ingenious theory of spectral lines which offered a solution of this difficulty. The theory is constructed on analogy with the fundamental note and various overtones of a vibrating string. These latter are harmonically related and analogously the wavelengths of the radiations emitted by a vibrating molecule are said also to be harmonically related. It would seem that Stoney may well have been influenced by the views of Maxwell, mentioned in section 1 of chapter II.[19]

It may not be assumed, argues Stoney, that the waves "impressed upon" the ether by one of the periodic motions of a gaseous molecule can be represented by a simple sine function.[20] They must be expected to be much more involved. Yet whatever the intricacy of their form at their source, the waves will retain substantially the same complex character while traversing the undispersing ether, in which waves of all frequencies travel at the same rate.

Mathematically speaking, the waveform, whatever its nature may be—whether one continuous curve, or parts of several different curves joined to one another, but assuming it to be plane—may be represented according to Fourier's theorem as a series of sines and cosines:

$$y = A_0 + A_1 \cos x + A_2 \cos 2x + \ldots$$
$$+ B_1 \sin x + B_2 \sin 2x + \ldots$$

which alternatively may be expressed in the form:

$$y - A_0 = C_1 \sin(x + a_1) + C_2 \sin(2x + a_2) + \ldots \qquad (1)$$

Now, says Stoney, the first term on the right-hand side of expression (1) represents a simple harmonic motion of period T, say; the remaining terms represent harmonics of this vibration, that is, their periods are $T/2$, $T/3$, etc. Thus expression (1) is equivalent to the statement that whatever be the form of the plane undulation it may be regarded as being formed by the superposition of a number of simple harmonic vibrations all harmonically related.

Stoney then proceeds to interpret this purely mathematical and hypothetical reasoning in physical terms. He first states that the fundamental

[19] See back to p. 43.
[20] G. J. Stoney, "On the cause of the interrupted spectra of gases," *Phil. Mag.*, 41 (1871): 291–96; p. 291.

vibration and the various harmonics will coexist in a state of mechanical independence of one another, provided the disturbance is not too violent for the legitimate employment of the principle of superposition of small motions. Furthermore, so long as the light traverses undispersing space these constituent vibrations will strictly accompany one another, since, as already mentioned, waves of all frequencies travel at the same velocity in open space. However, when the undulation enters a medium, such as glass, in which waves of different frequency travel at different rates, the constituent vibrations are no longer able to keep together, each being forced to advance through the glass at a speed depending on its frequency. Thus, there is a resolution of the original wave into its constituent components as represented by expression (1). And if the glass is in the form of a prism the successive components will emerge in different directions, so that each will give rise to a separate line in the spectrum of the gas.

Stoney admits that other expansions similar to Fourier's are possible, in which the terms, instead of representing simple harmonic vibrations, would represent vibrations of any other prescribed form; and hence it may be questioned whether the resolution effected by the prism is into components corresponding to the simpler series. While it cannot be directly demonstrated that the latter series actually occurs, Stoney advances considerations which seem to him to show his assumption "to be probable in so high a degree that it is the hypothesis which we ought provisionally to accept."[21] First, the arrangement of the emerging vibrations is independent of the material of the prism; and second, the arrangement is independent of the amplitude of vibration within very wide limits, since the positions of the lines remain fixed through great ranges of temperature. The first consideration, says Stoney, shows the series to be the same under varying circumstances; and the second suggests, as in the theory of superposition of small motions, that this series is one of simple harmonic vibrations.

Thus it is found "that one periodic motion in the molecules of an incandescent gas may be the source of a whole series of lines in the spectrum of the gas."[22] This is an ingenious way to avoid the difficulty of explaining how an iron molecule, for instance, can vibrate in a multitude of different ways on the supposition that each line in its spectrum corresponds to a distinct molecular vibration. Should some of the coefficients of series (1) vanish, then the corresponding lines will be absent from the spectrum. This is analogous, says Stoney, to the familiar case of the suppression of certain harmonics in music, and appears to

21 *Ibid.*, p. 293.
22 *Ibid.*

be what usually occurs in those spectra called by Plücker spectra of the second order (i.e., line spectra) .

Stoney thought that a gaseous molecule probably had several distinct motions, each giving rise to its own series of harmonics in the spectrum. In occurring together these series give the appearance of a confused maze of lines, as in line spectra. This apparent confusion seems greater due to the absence of most of the harmonics, so that it is not easy to trace the relationship between the few that remain.

Stoney applied his theory to the hydrogen spectrum consisting of the four lines corresponding to C, F, a line near G, and h in the solar spectrum. Of these he is able to ascribe three—C, F, and h—to a single internal motion of the gaseous hydrogen molecules. Taking Ångstrom's determinations of these lines, Stoney allows for dispersion in air and gives their wavelengths *in vacuo* as:

$$h = 4102.37 \times 10^{-10} \text{ meters,}$$
$$F = 4862.11 \times 10^{-10} \text{ meters,}$$
$$C = 6563.93 \times 10^{-10} \text{ meters.}$$

They are respectively the thirty-second, twenty-seventh, and twentieth harmonics of a "fundamental vibration" whose wavelength *in vacuo* is $131,277.14 \times 10^{-10}$ meters.[23] Stoney does not say how he found these harmonics, but one way would be to take each of the four lines in turn and draw up a column of the products of its wavelength with the integers from one to, say, fifty. Then it would be a matter of searching for the lowest product which is common to all four columns. If this indeed was Stoney's method he could only find a product common to three of the columns.

While the other harmonics of this "fundamental vibration" were not to be found in the ordinary spectrum of hydrogen, Stoney was hopeful that they might possibly be found in the other two spectra of hydrogen when the latter had been adequately mapped. This seems to be a careless statement by a man too eager to see that his theory is not weakened, for not only has he already explained that many harmonics may be suppressed but also (in 1868) that the different spectra of the one gas correspond to distinct molecular motions.

Stoney concludes that it is from the examination of band (first-order) spectra that the "most copious" results may be expected.

These spectra consist of lines ruled close to one another, and presenting in the aggregate the appearance of patterns which often resemble the flutings on a

---

23 For his fundamental vibration Stoney gives 0.13127714 millimeters which is equivalent to $1312771.4 \times 10^{-10}$ meters. He has made a slight slip.

pillar. When these spectra are more carefully examined, it is probable that the whole series of lines occasioning one of the fluted patterns will be found to be the successive harmonics of a single motion in the molecules of the gas. . . . But the necessary observations are at present almost altogether wanting. The only case in which the author has been able to arrive at any result was that of the nitrogen spectrum of the First Order, observed by Plücker. It would appear from his observations that the more refrangible of the two fluted patterns observed by him is due to a motion in the gas having a wavelength of about 0.89376 of a mm., which corresponds to a periodic time of 3×10^{-12} secs., one of the flutings consisting of the 35 harmonics from about the 1960th to the 1995th.[24]

But Stoney cautioned that this result was less to be relied upon than the one obtained with hydrogen. Stoney's treatment of nitrogen may be contrasted with that of Lecoq.

Wishing to make a satisfactory test of his theory Stoney together with J. Emerson Reynolds, Keeper of the Mineral Department and Analyst to the Royal Dublin Society, then investigated absorption spectra.[25] They sought a vapor whose absorption spectrum could be attributed to a single motion within the molecules and having lines sufficiently separated from one another to be easily measured. Such a vapor was that of chlorochromic anhydride. Of its 106 lines 31 were measured throughout its entire length.

In testing the theory against observation they found it convenient to use the reciprocals of the wavelengths, or inverse wavelengths. A scale of inverse wavelengths has the great advantage that a system of lines with periodic times that are harmonics of a given periodic time will be equally spaced upon it. Each number on Stoney's scale of inverse wavelengths could be regarded as the number of wavelengths in a millimeter.

Now if k is the inverse wavelength of a fundamental motion, its wavelength will be $1/k$th of a millimeter, and its harmonics will have wavelengths $1/2k$, $1/3k$, etc., millimeters. Accordingly the inverse wavelength of the nth harmonic will be $(n + 1)k$. Hence a system of harmonics which are equally spaced on an inverse wavelength scale at intervals of k divisions are harmonics of a fundamental motion whose inverse wavelength is k, and wavelength $1/k$th of a millimeter.

The inverse wavelength *in vacuo* was calculated for each of the thirty-one measured lines and compared with the corresponding inverse wavelength calculated on the hypothesis that the lines of the spectrum are equally spaced on the inverse wavelength scale, as they

[24] G. J. Stoney, "On the cause of the interrupted spectra of gases," p. 296.
[25] G. J. Stoney and J. E. Reynolds, "An inquiry into the cause of the interrupted spectra of gases, Part 2. On the absorption-spectrum of chlorochromic anhydride," *Phil. Mag.*, 42 (1871): 41–52.

ought to be according to the theory. To obtain the latter, calculated, inverse wavelengths, the difference k of any two consecutive measured inverse wavelengths was taken (no doubt they took an average value), then if the calculated inverse wavelength of the thirtieth line, say, is wanted, one simply adds thirty times this difference to the inverse wavelength of the first measured line.[26] The comparison of the two sets of inverse wavelengths showed that the differences of corresponding numbers fell within the permissible limits of error, and Stoney and Reynolds therefore concluded that in the case of chlorochromic anhydride Stoney's theory had been verified.

To J. L. Soret, editor of *Archives des Sciences Physiques et Naturelles,* the "extreme" precision with which Stoney's calculated wavelengths coincided with the observed values inspired "great confidence." Yet the high order of the harmonics, particularly for chlorochromic anhydride (from 628 to 733), as well as the absence of the greater number of the harmonics in the case of hydrogen, gave rise to doubts about the correctness of Stoney's hypothesis. However, "the coincidence of the calculated with the observed values is too exact for it to be possible to attribute it to chance: if not due to the existence of harmonics, it must proceed from some other determinate cause."[27] It seemed then to Soret that this "interesting" subject should be further studied.

Making use of Mascart's measurements of ultraviolet spectra, Soret quickly found several relationships "in a very superficial and incomplete examination of the question." Thus, taking the least refrangible lines of each of the three magnesium triplets, beginning with the least refrangible triplet, and taking the ratio of the wavelengths of the first two of these, Soret finds that it

is nearly identical with the ratio between the wavelengths of the hydrogen lines *C* and *F,* lines which Mr. Stoney regards as the 20th and 27th harmonics of one and the same fundamental vibration. The first two groups of magnesium also might therefore be regarded as the 20th and 27th harmonics of a fundamental group of vibrations, of which the wavelength, for the least refrangible line, would be 0.0103660 millim. As to the third group, it would not represent the 32nd harmonic (as is done by the hydrogen-line '*h*'), but, very nearly, the 31st.[28]

Similar relationships were also found in Mascart's cadmium spectrum. Thus the ratio of the wavelengths of the first and eighteenth

---

[26] Note, however, that in Stoney and Reynolds' paper the first line is numbered zero and not one; and that the first calculated inverse wavelength is not the same as the first measured one, as it should be if the calculation is carried out as I have indicated.
[27] J. L. Soret, "On harmonic ratios in spectra," *Phil. Mag.,* 42 (1871): 464–65.
[28] *Ibid.,* p. 465.

lines is "exactly" 5/2, and of the second and eighth lines 27/20. With respect to the latter two lines, and with Stoney's hydrogen results in mind, Soret sought but was unable to find the thirty-second harmonic. However, as in the case of magnesium, the thirty-first harmonic nearly coincides with the tenth line. The sixth line represents "very exactly" the twenty-third harmonic of the same fundamental. However, the sixth and tenth lines also form the same ratio of 27/20.

To Soret it seemed difficult to admit that all of these "coincidences" were fortuitous, and he thought it probable that others would be discovered in a closer examination of the question. But this task he left to other investigators. However, while in Soret's eyes Stoney's results inspired great confidence they do not seem to have aroused much general interest, at least for the time being. Perhaps Stoney himself was satisfied that he had vindicated his theory, for he did not return to it again until 1880.

3. An Important Photographic Development

Meanwhile, significant experimental advances were being made in the investigation of the ultraviolet parts of spectra, and results were obtained which were indispensable to success in the search for spectral series formulae. In this section I shall outline how the study of ultraviolet spectra was begun and then relate how it became relevant to the search for the hydrogen series formula. Its importance for the discovery of other series formulae will be seen in a later section.

Photography was first used in spectroscopy in the early 1840s when E. Becquerel and J. W. Draper each photographed the solar spectrum. But it was the early 1860s before its value was fully appreciated, and then primarily because it enabled spectroscopists to explore beyond the visible spectrum into the ultraviolet. This had also been made possible by the independent discoveries of E. Becquerel and G. G. Stokes, that unlike glass quartz has a very low absorption coefficient for the ultraviolet rays.[29]

Stokes reported in 1852 that in using a quartz prism he had found the light produced by a strong electric spark to be "excessively rich" in ultraviolet rays.[30] Ten years later he made a study of the ultraviolet spectra of metals using a prism and lens of quartz.[31] The faces of the prism were equally inclined to the axis of the quartz crystal from which

---

29 G. G. Stokes, "On the change of refrangibility of light," *Phil. Trans.*, 142 (1852): 463–562; p. 540.
30 *Ibid.*, p. 546.
31 G. G. Stokes, "On the long spectrum of electric light," *Phil. Trans.*, 152 (1862): 599–619.

it was cut, and the lens was cut perpendicular to the axis. Stokes rendered his spectra visible by projecting them on to a fluorescent screen.

In the same year (1862) ultraviolet metallic spectra were also studied by William Allen Miller, who instead of using a fluorescent screen photographed his spectra.

The lines of each spectrum are so numerous and so close together, that it would be impossible without a sacrifice of time, that would scarcely be justifiable, to obtain accurate impressions of them by eye-drawing. Indeed, except by the process of photography, these lines can only be rendered visible by the aid of a fluorescent screen, under which circumstances the minute details are almost necessarily lost even by the most careful observer.[32]

Miller photographed the ultraviolet spectra of some thirty metals. They were "remarkable" in that they extended beyond the visible spectrum as much as "five or six" times the length of the latter. While the lines became progressively less well defined with increase of refrangibility, as can be seen from the plates published with the paper, Miller observed a certain similarity in the spectra of "allied" metals, as in the case of cadmium, zinc, and magnesium; iron, cobalt, and nickel; bismuth and antimony; and chromium and manganese. And this was several years before Lecoq's work appeared.

In Miller's apparatus the telescope of the usual spectroscopic arrangement is replaced by a camera. A wet collodion plate was used. Miller complained of the comparatively low dispersion of quartz and of the disadvantage of the effects of double refraction, but he could find no preferable medium. He positioned his prism so as to obtain a singly refracted beam for the intermediate wavelengths.

In 1867 Mascart rendered valuable service to spectroscopists when he photographed and accurately measured the ultraviolet lines of cadmium.[33] Three years earlier he had done equally excellent work in photographing and measuring the ultraviolet lines of the solar spectrum with a thoroughness and accuracy similar to that achieved by Kirchhoff in the visible solar spectrum.[34] His method of photography was, however, most laborious.

During the 1870s a happy development for spectroscopists and

[32] W. A. Miller, "On the photographic transparency of various bodies, and on the photographic effects of metallic and other spectra obtained by means of the electric spark," *Phil. Trans.*, 152 (1862): 861–87; p. 876.

[33] E. Mascart, "Recherches sur la détermination des longuers d'onde," *Annales Scientifiques de L'École Normale Supérieure*, 4 (1867): 1–31.

[34] E. Mascart, "Recherches sur le spectre solaire ultra-violet et sur la détermination des longuers d'onde," *Annales Scientifiques . . .* , 1 (1864): 219–62.

photographers alike occurred within photography. This was the invention and large-scale production of the "dry plate."[35]

The disadvantages of the "wet" process such as had been employed earlier by Miller, when compared with the "dry" process, were described jointly in 1879 by Walter Noel Hartley and his colleague A. K. Huntington.[36] The disadvantages were three in number. As already noted, the more refrangible end of the ultraviolet spectrum was extremely weak, if not entirely wanting. Second, the wet process was troublesome when long exposures were required. Finally, when working in a small room the ozone generated by the electric discharge acted on the wet collodion plates in such a way that they became coated with a thick deposit of silver when the developing solution was applied. However, the "extraordinary improvements made of late years in the preparation and development of dry plates" apparently allowed all of these disadvantages to be overcome. One could choose the Beechy standard dry plates or the gelatin pellicle plates of either Kennett or Wratten and Wainwright, but on the whole the gelatin plates were to be preferred.

The dry plate was likewise a welcome addition to the apparatus of solar and stellar spectroscopy. This is an area which was intensively investigated by such men as Ångstrom, Kirchhoff, Van der Willigen, and Mascart. For the most part it lies beyond our direct interest, but at one point stellar studies yielded some important data in the quest for spectral series formulae.

In the late 1870s the astronomer William Huggins found for himself the advantages of using dry plates.

[First] I used wet collodion, but I soon found how great would be the advantages of using dry plates. Dry plates are not only more convenient for astronomical work, being always ready for use, but they possess the great superiority of not being liable to stain from draining and partial drying of the plates during the long exposures which are necessary even with the most sensitive plates. I then tried various forms of collodion emulsion, but finally gave up in favour of gelatine plates, which can be made more sensitive.[37]

Huggins's plates measured $1\frac{1}{2} \times \frac{1}{2}$ inches, and the definition was so good that the photographs could be examined with advantage under a microscope.

35 See "The evolution of dry plates," in H. and A. Gernsheim, *The History of Photography* (London, 1955), pp. 243–54.

36 W. N. Hartley and A. K. Huntington, "Researches on the action of organic substances on the ultra-violet rays of the spectrum," *Phil. Trans.*, 170 (1879), part 1, pp. 257–74; p. 260.

37 W. Huggins, "On the photographic spectra of stars," *Phil. Trans.*, 171 (1880): 669–90; p. 673.

As Iceland spar has a higher dispersive power than quartz, and as the number of prisms used should be kept to a minimum because of the weak intensity of starlight, Huggins worked with a single prism of Iceland spar. It was cut in a plane perpendicular to the axis of the crystal and thus in any one position gave single refraction for one wavelength only. However, it was found in practice that the separation of the ordinary and extraordinary rays throughout the range of the spectrum photographed was too small sensibly to affect the results.

Huggins found the spectra of certain white stars to possess in common twelve "very strong" lines. The two least refrangible lines coincided with the hydrogen lines γ and h and the third with the solar line H; but the other nine more refrangible lines did not appear to be coincident with any of the stronger solar lines. Huggins drew attention to the "remarkable arrangement" of all twelve lines. With increase in refrangibility the breadth of the lines as well as the distance between any two adjacent lines decreases. "The group possesses a distinctly symmetrical character. The suggestion presents itself whether these lines are not intimately connected with each other, and present the spectrum of one substance."[38] The substance which Huggins had in mind was hydrogen, a probability which had the support of two of his scientific acquaintances, H. W. Vogel and Stoney.

In a letter to Huggins, Vogel pointed out that four additional lines which he had found with hydrogen agreed with the third, fourth, fifth, and sixth of Huggins's white star lines.[39] Vogel had found these and other lines in the ultraviolet to be given by hydrogen excited by the electrical discharge in a Geissler tube.[40] He had been able to do so only with the help of the dry plate. At first he had tried to use wet plates but found that in addition to not being sensitive enough they also dried too quickly to permit the relatively weak light of a Geissler tube to be photographed. However, he found the gelatin dry plates to have a "wholly uncommon" sensitivity—more than fifteen times as great as that of wet plates. Hence the spectrum of the weak light of a Geissler tube, just as the weak light of a star, could now be photographed. Curiously, Vogel used a flint-glass prism.

In another letter to Huggins, and apparently without knowledge of Vogel's work, Stoney advanced different evidence for considering

[38] *Ibid.*, p. 678.
[39] An extract of Vogel's letter is published in Huggins, "On the photographic spectra of stars," p. 678.
[40] H. W. Vogel, "Ueber die photographischen Aufnahme von Spectren in Geisslerröhren eingeschlossenen Gase," *Monatsberichte Akad. Wissen. Berlin* (1879): 115–19. "Ueber die Spectra des Wasserstoffs, Quecksilbers und Stickstoffs," *Monatsberichte . . . Berlin* (1879): 586–604.

Huggins's lines to be hydrogen lines.[41] The likelihood of their all, including the four known lines of hydrogen, being members of "one physical system" was made very plain, Stoney thought, when their frequencies were plotted as in Figure 6. For "it then becomes more conspicuous that they lie on, or very near, a definite curve, which would not happen by chance."[42]

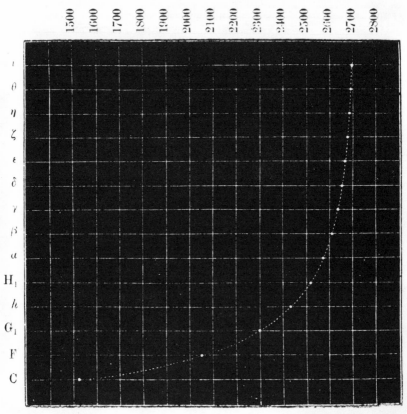

Fig. 6. Lockyer's Graphical Representation of White Star Lines

The question of whether they lie actually on, or only near, a definite curve was to Stoney of "very great" theoretical significance. If they lie on a curve obeying any exact mathematical law, their connection must, he thought, be attributed to their corresponding to

41 Stoney's letter is also published in Huggins, "On the photographic spectra of stars," pp. 678–81.
42 *Ibid.*, p. 679.

the *consecutive* partial tones of some vibrating system (like those of an elastic rod or bell for example). If, on the other hand, they lie near but not on the curve this circumstance would support the hypothesis (which seems to accord with other facts) that the visible lines are members of a harmonic series, most of the members of which are invisible, those only being seen whose positions chance nearly to fulfill a definite condition—a state of things which I have shown to exist in some acoustic arrangements, and which whenever it prevails exalts the intensity of the harmonics whose positions nearly fulfill the requisite condition.[43]

To decide between these two alternatives Stoney drew up a table in which the first column contains the wavelengths of Huggins's lines, including the four hydrogen lines; the second the corresponding frequencies of these wavelengths; the third the difference between consecutive frequencies; and the fourth the differences between successive figures of column three. And without any explanation to Huggins, Stoney concludes that "assuming that the irregularities in the second differences cannot be referred to errors of observation, I think that the accuracy of your work gives evidence which must be accepted that the second alternative is the true one, viz., that the lines do not lie on but near a definite curve."[44]

Stoney continues that, and this will become significant later, the table shows that the lines in question are not consecutive members of one series, but rather the members of one or more series whose positions lie near the curve. This appeared to be corroborated when it was found that H_1 and the hydrogen line $H\gamma$ are harmonically connected, their wavelengths being exactly the thirty-fifth and thirty-second harmonics respectively of a vibration whose period is $T/72.003$ (T being the time in which light travels one millimeter in air). This was a different series from the one which Stoney had previously established for the hydrogen lines C, F, and h. The remaining lines observed by Huggins did not belong to either of these series, and in order to fit them to his theory Stoney had to suppose "two other motions at least to exist in hydrogen." He suggested that possibly six of the remaining lines a, β, γ, δ, ξ, and η were respectively the 284th, 288th, 291st, 293rd, 296th, and 297th harmonics of the fundamental of period $T/9.0572$; and the final three lines ϵ, θ, and ι to be respectively the 390th, 394th, and 395th harmonics of the fundamental of period $T/6.845$. Thus while Stoney did "not attribute much weight to the last two series" he was nevertheless prepared to allow a hydrogen molecule four fundamental and unrelated

43 *Ibid.*
44 *Ibid.*, p. 680.

oscillatory motions. Still this was simpler than assuming a distinct motion of the molecule for each spectral line.

On the whole this work did little to strengthen Stoney's theory, which had been mildly criticized for the first time in the previous year (1879) by Arthur Schuster.[45]

4. Stoney's Theory Refuted

A further year passed apparently uneventfully, but all the while Schuster was bringing to completion a three-year study of Stoney's theory, and now, in 1881, he published his results. For the most part they pointed to the conclusion that Stoney's theory was untenable; and from this time on the search for simple harmonic series was abandoned.

Schuster explains that in any spectrum containing a large number of lines (he does not say what is meant by "large," but as can be seen below he considers a spectrum with as few as seven lines) it is always possible, due to chance, to find lines the ratio of any two of whose wavelengths agrees very closely with the ratio of small integers (as can also be seen below, 100 is considered a small integer!).[46] However, by means of the theory of probability it is possible to calculate the number of such coincidences on the supposition that no law exists, and that the lines are distributed at random throughout the visible spectrum. If then the number of ratios of any two lines which agree within certain limits with ratios formed of integers "greatly exceeds" the most probable number, there is reason to suppose that the lines are not distributed at random, but that Stoney's law is true.

Schuster employs two methods and actually only uses random distribution calculations in the second. In the first he considers the spectra of magnesium, sodium, copper, barium, and iron, which according to Ångstrom have respectively 7, 10, 15, 26, and 149 lines. He compares the ratio of the wavelengths of any two lines in each of these spectra with ratios of integral numbers smaller than 100, calculated to six decimal places. For example, with sodium the ratio of the wavelength of the less refrangible of the two yellowish-green lines and that of the less refrangible of the two yellow lines comes to 0.964760. On referring to his table of ratios of integral numbers Schuster finds that the ratio of the sodium lines lies between

$$55 \div 57 = .964912$$

and $$82 \div 85 = .964706, \qquad (A)$$

45 A. Schuster, "On harmonic ratios in the spectra of gases," *Nature*, 20 (1879): 533.
46 A. Schuster, "On harmonic ratios in the spectra of gases," *Proc. Roy. Soc.*, 31 (1881): 337–47.

the difference of these being   .000206.   (B)

The difference between the ratio of the sodium lines and the closest lying ratio of integers (A) is .000054.   (C)

The ratio of the two differences (C) and (B) is

$$54 \div 206 = .262.$$

Ratios similar to the latter one were then found for all possible cases in the sodium spectrum.

Now, says Schuster, if the lines in spectra are distributed at random we should expect the ratio of the two differences to range indiscriminately between zero and 0.5; the mean of all of them coming near to 0.25. If, on the other hand, the law of harmonic ratios is true we should expect a greater number of small fractions, and hence the mean should be smaller than 0.25.

The results for all five spectra studied are as follows.

Spectrum	Mean value of ratio
Magnesium	.2626
Sodium	.2399
Copper	.2430
Barium	.2592
Iron	.2513
Mean	.2514

Schuster's method was sound and he correctly and enthusiastically concluded that nothing "could be more decisive against the law of harmonic ratios than this table; three out of the five elements considered, including the two containing the greatest number of lines, give a mean value greater than .25."[47]

Employing a "more direct and complete" method, suitable for use with spectra having large numbers of lines, Schuster then reconsidered the iron spectrum and found himself in somewhat of a dilemma. In this second method he does calculate how many pairs of lines "agree within certain small limits with harmonic ratios" assuming a random distribution of the lines, and then compares the number obtained with the number found in practice.

Schuster again uses Ångstrom's wavelength determinations and employs two different limits within which a harmonic relationship is considered to have occurred (see table). He explains that a decided advantage is gained in taking the results with two limits. "It is in fact

[47] *Ibid.*, p. 338.

equivalent to using a third method of discussion, for supposing the spectral lines to be distributed at random, the number of coincidences found should be proportional to the limits chosen. If, on the other hand, the law of harmonic ratios is correct, the narrower limit should relatively show the greater number of coincidences."[48]

It is unnecessary here to consider Schuster's derivation, given in an appendix, of the probable number of harmonic relationships to be expected in a random distribution of lines; but it should be noted that as he does not give the number, or wavelengths, of the iron lines employed his results could not be checked. They are as shown in the table.

	Limits ± .0000505		Limits ± .0000755	
	Observed	Calculated	Observed	Calculated
0–10	48	52	64	77
10–20	180	206	250	308
20–30	329	363	469	544
30–40	478	521	664	779
40–50	625	679	912	1,015
50–60	777	837	1,163	1,251
60–70	886	968	1,318	1,447
70–80	924	896	1,337	1,340
80–90	667	629	989	940
90–100	253	241	393	361
Total	5,167	5,392	7,559	8,062

For all wavelength ratios with denominator less than seventy the calculated coincidences are in excess of the observed ones and so contest the theory of harmonic ratios. However, the situation is reversed for all ratios with denominator lying between 70 and 100. Furthermore, the smaller limit gives results decidedly more favorable to the theory than the larger limit. This fact, even though the observed number of coincidences is less than the calculated number for both limits, suggested the possibility that still narrower limits might give results which are yet more favorable to the theory of simple harmonic ratios. This indeed is the case, and Schuster was faced with a dilemma. Had this investigation of the iron spectrum confirmed or refuted the theory of harmonic relationships?

Schuster considered his findings:

1. *There is a real cause acting in a direction opposed to the laws of harmonic ratios, so far as fractions formed by numbers smaller than seventy are concerned.*

[48] *Ibid.*, p. 339.

2. *After elimination of the first cause a tendency appears for fractions formed by two lines to cluster round harmonic ratios.*

And then he concluded that:

3. *Most probably some law hitherto undiscovered exists, which in special cases resolves itself into the law of harmonic ratios.*[49]

Thus while the results obtained by the first method decisively contradicted the theory of harmonic relations, the second method applied to the iron spectrum gave no such definite result. With regard to the latter spectrum Schuster considered further investigation desirable. One might be able to confirm the laws found with the iron spectrum in investigating other spectra having many lines, as those of manganese or calcium. But Schuster thought it "more promising to increase the accuracy of measurement in the special cases where harmonic relations have been found." And to this end he appended a list of pairs of iron lines which are "nearly" in the ratio of some fraction formed by integers smaller than ten (as 2/3, 3/4, 4/5, 5/6, 6/7, 9/10).

Although Schuster could not decisively refute the theory of simple harmonic relations, the effect of his work was to discourage others from seeking such relations.[50] Schuster himself did not pursue the question further. While he had "no doubt" that a law relating spectral lines did exist, he was quite at a loss to suggest how it might be found. He had overturned Stoney's theory, but he had not replaced it by another. Nor was one immediately forthcoming from another quarter.

Schuster was one of the leading British spectroscopists and, as seen in chapter II, had prepared a progress report on spectroscopy for the meeting of the British Association for the Advancement of Science, in 1880. Two years later he prepared a second report, this time on "molecular vibrations." Regarding the search for numerical relationships among spectral lines he refers to his criticism of Stoney's theory and then for the first time explicitly states what had been implicit in Stoney's work. "It is the ambitious object of spectroscopy to study the vibrations of atoms and molecules in order to obtain what information we can about the nature of the forces which bind them together."[51] However, ambition was not to be coupled with any great optimism—

But we must not too soon expect the discovery of any grand and very general law, for the constitution of what we call a molecule is no doubt a very compli-

---

[49] *Ibid.*, p. 343.
[50] See report of 12 June, 1885, meeting of Berlin Physical Society in *Nature*, 32 (1885): 312. It is stated here that the attempt to find simple harmonic relations was abandoned after the question had been discussed by Schuster.
[51] A. Schuster, "The genesis of spectra," *Rep. British Assn.*, 50 (1882): 120–43; p. 120.

cated one, and the difficulty of the problem is so great that were it not for the primary importance of the result which we may finally hope to obtain, all but the most sanguine might well be discouraged to engage in an enquiry which, even after many years of work, may turn out to have been fruitless. We know a great deal more about the forces which produce the vibrations of sound than about those which produce the vibrations of light. To find out the different tunes sent out by a vibrating system is a problem which may or may not be solvable in certain special cases, but it would baffle the most skilful mathematician to solve the inverse problem and to find out the shape of a bell by means of the sounds which it is capable of sending out. And this is the problem which ultimately spectroscopy hopes to solve in the case of light. In the meantime we must welcome with delight even the smallest step in the desired direction.[52]

There can be little question that this pessimistic attitude was typical of the time. Schuster himself turned away from spectroscopy and sought "the shape of the bell" in another direction—the study of the passage of electricity through rarified gases, to which he also made significant contributions.[53]

In the remainder of his relatively long report Schuster did not say much that was favorable to earlier work. Lecoq's investigation of harmonic relationships and Stoney's work on chlorochromic anhydride did not have the desired accuracy which was now possible.[54]

However, on the positive side we do find the recurring idea that laws relating spectral lines are to be found. Schuster mentions the repeated doublets, triplets, and quadruplets, which were being widely recognized in spectra now that the ultraviolet was being studied. In this regard the highly valued experimental work of the Cambridge team of George D. Liveing and James Dewar was most important. They drew attention in 1879 to quadruplets in the sodium and potassium spectra—"the repetition of these quartets of lines at decreasing intervals with decreasing brightness and sharpness as they proceed from the less to the more refrangible, gives the impression of a series of harmonics; but the wavelengths do not seem to be in a simple harmonic progression, though simple harmonic relations may be found between some of the groups."[55] In the following year they found the lithium lines to form a progression "not unlike" that of the sodium lines.[56] And

52 *Ibid.*, pp. 120–21.
53 See: David L. Anderson, *The Discovery of the Electron* (Princeton, 1964), pp. 30–32; E. Whittaker, *A History of the Theories of Aether and Electricity* (New York, 1960), 1: 355–56.
54 A. Schuster, "The genesis of spectra," pp. 121–22.
55 G. D. Liveing and J. Dewar, "On the spectra of sodium and potassium," *Proc. Roy. Soc.*, 29 (1879): 398 ff.; in G. D. Liveing and Sir J. Dewar, *Collected Papers on Spectroscopy* (Cambridge, 1915), pp. 66–70; p. 70.
56 G. D. Liveing and J. Dewar, "On the spectra of magnesium and lithium," *Proc. Roy. Soc.*, 30 (1880): 93 ff.; in *Collected Papers*, pp. 78–84; p. 84.

in 1881, series of pairs and triplets in the magnesium spectrum were also likened "in general character" to the sodium and potassium series.[57] Schuster thinks that this is "very suggestive, and promises to furnish a safer basis for future research than the hypothesis of [simple] harmonic relationships."[58] But he fails to elaborate.

Schuster's intuition was, nevertheless, in part, correct, for in continuation of their previous work a paper published in 1883 by Liveing and Dewar was eventually to prove most important. This contained the first results of a three-year investigation of the ultraviolet spectra of fifteen metals in which "some thousands" of photographs had been examined. Liveing and Dewar note that they have previously called attention to probable harmonic relations in the individual spectra of the alkali metals and magnesium.

This relation manifests itself in three ways—first, by the repetition of similar groups of lines; secondly, by a law of sequence in distance, producing a diminishing distance between successive repetitions of the same group as they decrease in wavelength; and thirdly, a law of sequence as regards quality, an alteration of sharper and more diffuse groups, with a gradually increasing diffuseness and diminishing intensity of all related groups as the wavelength diminishes.[59]

For example, lithium is found to have two harmonic series of single lines. Sodium and potassium have each a harmonic series of doublets and one of quadruplets; and calcium and zinc have series of triplets while aluminum has doublets and triplets. By "harmonic series" Liveing and Dewar merely mean a series of overtones of a fundamental vibration—"we do not mean that they follow the simple arithmetical law of an ordinary harmonic progression, but are comparable rather with the overtones of a bar or bell than with those of a uniform stretched string."[60] The alternations of sharper and more diffuse groups are generally apparent and are very marked in the spectra of calcium and zinc. The decrease of separation and of intensity and the increase in diffuseness of successive members of the series toward smaller wavelength is clearly seen in all the cases mentioned.

Knowledge of series was further advanced in the same year by Hartley, who found that the separation of the two components of a doublet

[57] G. D. Liveing and J. Dewar, "Investigations on the spectrum of magnesium. No. 1," *Proc. Roy. Soc.*, 32 (1881): 189 ff.; in *Collected Papers*, pp. 118–32; p. 120.
[58] A. Schuster, "The genesis of spectra," p. 139.
[59] G. D. Liveing and J. Dewar, "On the ultra-violet spectra of the elements. Part 1. Iron (with a map). Part 2. Various elements other than iron (K, Na, Li, Ba, Sr, Ca, Zn, Au, Tl, Al, Pb, Sn, Sb, Bi, Si, C)," *Phil. Trans.*, 174 (1883): 187 ff.; in *Collected Papers*, pp. 193–230; p. 221.
[60] *Ibid.*, p. 222.

(and the separations of the three components of a triplet) when measured in wave numbers, was a constant throughout a given series.[61]

Two years were to pass before the next step was taken in the logical extension of these discoveries of Liveing, Dewar, and Hartley. In the meantime no one advanced a hypothesis to replace Stoney's, although in redefining "harmonic series" Liveing and Dewar had given a generalized and qualitative version of Stoney's original hypothesis. Indeed, in 1885 the use of hypotheses was condemned by Alfred Cornu, Professor of Physics at the École Polytechnique in Paris, who at the same time took the step referred to.

Cornu reiterates the view that the search for simple harmonic ratios seems to offer no hope, but continues:

> This lack of success ought not to discourage observers; still the hope of finding a simple law, as that of musical harmonics, is the sign of a preconceived idea which it is important to discard immediately: this law of whole numbers applies only to a very particular kind of sonorous body of which the type is the cylindrical column of great length in relation to section: if the form of the vibrating body departs from this special type, the relation between the frequencies of successive sounds becomes very complex.[62]

Cornu thought it necessary to reject all preconceived ideas on the mathematical relationships of the phenomena and to investigate experimentally whether a common law could not be found for the many series whose regularity and similarities are "evidently not fortuitous." In studying visible and ultraviolet spectra Cornu himself had found such a law for "spontaneously reversible" lines—which he had previously shown to occur in the spectra of most metallic vapors.

Cornu explains that at low temperature and pressure an incandescent vapor emits certain narrow bright lines. As the temperature and pressure are steadily increased the lines at first increase in width and intensity to become luminous bands with blurred edges. But after a time dark lines begin to appear in the bands in the positions of the original bright lines. These dark lines are the spontaneously reversible lines and they are absorption phenomena. The bright nonreversible lines usually disappear completely as the bright bands become more diffuse.

Cornu finds that in most spectra spontaneously reversible lines are "reproduced periodically with a particular regularity." Furthermore, *"the lines draw closer together towards the more refrangible side and*

---

[61] W. N. Hartley, "On homologous spectra," *J. Chem. Soc. London,* 43 (1883): 390–400.
[62] A. Cornu, "Sur les raies spectrales spontanément renversables et l'analogie de leur lois de répartition et d'intensité avec celles des raies de l'hydrogène," *Comptes Rendus,* 100 (1885): 1181–88; p. 1182.

at the same time diminish in intensity." The most beautiful series observed by Cornu are given by two metals "which one would hardly expect, from the chemical point of view, to find side by side: these are aluminum and thallium whose equivalents are at the extremities of the list of those of simple bodies." Each is a series of doublets and a remarkable correspondence in structure is found between each of them and the hydrogen spectrum of Huggins (see Fig. 7). In each the first lines of the doublets form one series, and the second lines another series. This is the natural extension of the work of Hartley and Liveing and Dewar. All four series of single lines are quite similar to the hydrogen series. In fact the wavelengths of the two aluminum series are respectively given by

$$\lambda_1 = 47.30 + 0.43783h$$
$$\lambda_2 = 47.18 + 0.43678h$$

and those of the thallium series by

$$\lambda_1 = 94.61 + 0.29776h$$
$$\lambda_2 = 111.31 + 0.75294h,$$

where h is the wavelength of the corresponding hydrogen line.

Cornu also investigated more complex spectra such as those of magnesium, zinc, and sodium, and his general conclusion was that in *"metallic spectra certain series of spontaneously reversible lines display appreciably the same laws of distribution and intensity as the hydrogen lines."* It is not necessary, he says, to insist at length on the importance of this relation. On the one hand the existence of a general law governing the emissive powers of incandescent vapors is revealed. On the other it is seen that the same function, which one might call the "fonction hydrogènique," governs the distribution of lines in all the series. To Cornu the preceding results appeared to be a first step toward the solution of the "great problems which are posed in spectroscopy." We shall see that they must certainly have been highly suggestive to certain other investigators who were later successful in finding satisfactory spectral series formulae. However, with more accurate spectral determinations the factor B in Cornu's own expression $\lambda = A + Bh$ was found not to be a constant.[63]

It is of interest to note that it was the "beautiful" discovery of the hydrogen spectrum by Huggins which had encouraged Cornu to continue his researches—after he had attempted unsuccessfully to represent the lines of a series by a simple function in which the series of

[63] See H. Kayser, *Handbuch der Spectroscopie* (Leipzig, 1902), 2: 507.

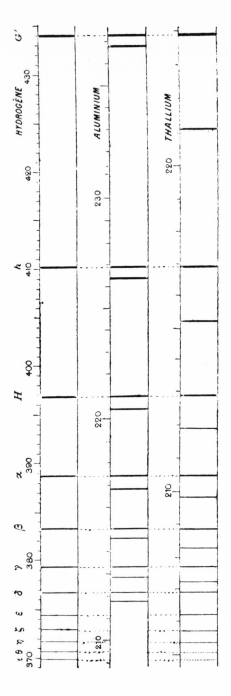

Fig. 7. Drawings of the Hydrogen, Aluminum, and Thallium Spectra

integers would be successively substituted. In contrast both to Cornu's failure in this latter respect and to his consequent empirical approach, an obscure sixty-year-old Swiss schoolmaster had, unknown to Cornu, just a few months earlier published a remarkable paper—his first on spectra.

5. Balmer's Discovery of the Hydrogen Series Formula

His published papers do not reveal just how long Johan Jacob Balmer, a mathematics teacher, had searched for numerical spectral relationships before discovering the formula for the hydrogen series.[64] It may have been weeks, months, or even years. Again it cannot be said what numerical methods he employed in his computations or just which parts of the work of his contemporaries were known to him.

Certainly Balmer knew of Stoney's earlier work, for he explains that the simple whole-number relationships shown by Stoney to exist between three of the hydrogen lines had suggested an acoustical analogy —that the spectral lines of a given substance might be conceived as the overtones of the one "key-note." He adds, however, that none of the attempts to find such a key-note had produced satisfactory results. But contrary to what we might think, Balmer does not appear to be following Schuster here. Rather, these attempts led to such great numbers "that therewith a clearer insight was not achieved."[65] For example, for the first, second, and fourth hydrogen lines Stoney obtained a key-note which was twenty-seven times as long as the wavelength of the second line. But what precisely was Balmer's objection? Was it that such a line had not been seen and had little prospect of being seen? That is, did he regard the key-note as being unreal? Perhaps that is the answer, and perhaps not. When he says that a clearer insight has not been obtained we might think that Balmer has in mind Stoney's four independent motions of the hydrogen molecule; but as we shall see Balmer did not know of this part of Stoney's work until after he had arrived at his formula. One cannot help feeling that being interested in spectral series Balmer must have known of Schuster's work, yet he neither mentions Schuster nor indicates that the search for harmonic relationships is to be abandoned.

[64] J. J. Balmer, "Notiz über die Spectrallinien des Wasserstoffs," *Verhandlungen der Naturforschenden Gesellschaft in Basel*, 7 (1885): 548–60, 750–52. (A shorter version of these two papers also appears in *Ann. der Physik*, 25 (1885): 80–87.) In reconstructing the steps of Balmer's discovery I have also made use of a biographical note by A. Hagenbach published in *Die Naturwissenschaften*, 9 (1921): 451–53. This Hagenbach is a son of the one mentioned in the text.
[65] J. J. Balmer, "Notiz über die Spectrallinien des Wasserstoffs," p. 552.

The simple numerical relationships mentioned above had convinced Balmer that some simple law existed which would describe the hydrogen lines. In seeking this law he had the encouragement of a colleague named Hagenbach, with whom he often discussed scientific and educational matters. One day Balmer confided to Hagenbach that he could relate the three hydrogen lines H_α, H_β, and H_γ, to a single "key-note" by means of proper fractions. A short time later Balmer was able also to include the fourth hydrogen line H_δ. The four proper fractions are 9/5, 4/3, 25/21, and 9/8, and the key-note or "fundamental number of hydrogen" is given by $h = 3645.6 \times 10^{-7}$ millimeters.

Stoney had only been able to relate three of the hydrogen lines to a fundamental, but now Balmer had related all four. Furthermore, the hydrogen line of longest wavelength is not quite twice that of the fundamental number, and no doubt Balmer, for whatever reason, derived satisfaction from this. The fractions and the fundamental number express the four hydrogen wavelengths to a high degree of precision. But the fractions are not those of a harmonic series—the "overtones" in this case are all greater in wavelength than the fundamental, whereas in a harmonic series of the acoustical type they are smaller. Indeed the fractions do not even form a series.

However, the basic belief of the scientist is that natural phenomena are ordered and therefore knowable. And it was Balmer's strong conviction that a law existed. Unfortunately he does not tell us how long he struggled or what strategies he employed, but eventually he discovered that as soon as one multiplies the numerator and denominator of the second and fourth fractions by four the law is apparent: 9/5, 16/12, 25/21, 36/32. The series is given by $m^2/(m^2-n^2)$ where $n = 2$ and m is the series of integers beginning with three.

Balmer's joy was great. However, a fifth hydrogen line predicted by the formula ought to lie in the visible region close to the solar line H measured by Ångstrom, yet Balmer knew of no such line. Either the temperature of the sun was unfavorable to the production of this line, or his formula possessed no "universal value." Balmer decided to tell Hagenbach of his dilemma. Happily the latter knew of the work of Huggins and Vogel and was able to test Balmer's formula against their determinations of the ultraviolet hydrogen lines. In all cases the formula gave slightly greater values than the observed ones, but the differences were so small that "the agreement must astonish in the highest degree."[66] One can imagine Balmer's elation upon receiv-

66 J. J. Balmer, *Ann. der Physik*, 25 (1885): 80–87; p. 82.

ing Hagenbach's reply. The latter promptly had a notice of the discovery published.[67]

While Balmer had not found a simple harmonic relationship such as that associated with a vibrating string, it is nevertheless clear that he had been greatly assisted by Stoney's hypothetical analogy. For some reason he had modified Stoney's approach and sought a much smaller number for his fundamental. But the analogy remained essentially intact and proved indispensable to the discovery of the series formula. One wonders what thoughts passed through Cornu's mind when he eventually read Balmer's paper. Certainly the discovery came as a surprise to all spectroscopists.

6. Discovery of Series Formulae for Band Spectra

Meanwhile a significant beginning was being made in understanding the structure of band spectra, although the first results were not published until 1886. As we shall see, the necessary data for this, as indeed for most of the spectral series investigations, was made available by persons who were not at all interested in finding series formulae but rather in adding to astrophysical knowledge. Astrophysics was undoubtedly the major spectroscopic field in the second half of the nineteenth century, and most spectroscopists—including Kirchhoff, Ångstrom, Cornu, and Rowland, all of whom produced maps of the solar spectrum—had some active interest in the subject. Its literature was enormous, although it was not until the last decade of the century that the *Astrophysical Journal* appeared. It was because of this great interest in astrophysics that abundant spectroscopic materials were readily available for a Balmer or a Rydberg, whose tools were pencil and paper rather than spectroscope and photographic plate. Furthermore, it is the astrophysical interest which partly explains the largely fact-gathering investigations of such eminent and respected experimental spectroscopists as Liveing and Dewar. Knowledge of the chemical composition and physical condition of heavenly bodies required a prior and thoroughly detailed knowledge not only of the latter's spectra but also of the spectra of terrestrial elements. It was a laborious and unglamourous task, although by no means a unique one in astronomy. Fortunately, however, there were some who tackled it.

Such a person was C. Piazzi Smyth, Astronomer Royal for Scotland, who in the early 1880s supplemented the work of Ångstrom and Thalén on the spectra of gases by describing the spectra of additional gases

[67] In *Verhandlungen der Naturforschenden Gesellschaft in Basel,* for April, 1884.

examined under very high dispersion. In the course of this work Smyth had drawn a highly detailed map of the green carbon monoxide band, resolving many lines previously regarded as broad and even fine into doublets and triplets.[68] He forwarded a copy of this to Alexander S. Herschel, second son of John Herschel and Professor of Physics at the Durham College of Science in Newcastle, who compared it with a map of the same band which he himself had made in the previous year (1882). In an animated letter to Smyth, Herschel writes:

[I] couldn't make them fit immediately, until I found that you had duplicated and triplicated numbers of the lines that I recorded "broad," "winged," "united pair," etc., only, so that there is a profusion of new dissections of the band that you have managed now to supply for its anatomy! And then Io Triumphe! In searching over the spaces of my "readings" to identify your lines with, I lighted luckily *on the key* of the construction, which is simplicity itself, and couldn't well be exceeded in the exactness with which your new map reveals it! *Lux in tenebris,* what a happy and glorious release you have disclosed to all our uncertainties![69]

Herschel first observed that while the separation of successive "leaders" and also of successive "twin-cub followers" (see Fig. 8) increased regularly in one direction, it was not so with the separations

Fig. 8. *Drawing of the Green Carbon Monoxide Band*

of twin-cubs and the leaders next following them. In this latter case the distance a remains constant. Then upon examining the entire band Herschel found to his "joyful surprise" that it was *"simply two exactly similar* single-rank line progressions laid over each other displacing one of them slightly on the other; and while one consists of single strong lines, the other is formed of fainter, closely double ones." Several months later, in 1884, Herschel wrote again to Smyth to say that if the first members of each of the two series were made to coincide then an approximate coincidence would be found between the other corresponding lines—the single lines coinciding with either of the two doublet components or falling between them.[70] Furthermore, as measured on his arbitrary scale, the separations of successive members, be-

68 C. P. Smyth, "Micrometrical measures of gaseous spectra under high dispersion," *Trans. Roy. Soc. Edinburgh,* 32 (1887): 415–60.
69 *Ibid.,* Appendix, pp. 454–56.
70 *Ibid.,* p. 457.

ginning from the least refrangible member, were as the summation of integers—

$$1 : 1 + 2 : 1 + 2 + 3 : \ldots$$
$$1 : \quad 3 \quad : \quad 6 \quad : \ldots$$

Although Herschel's letters were written in 1883 and 1884 they were published, as appendices to Smyth's work, only in 1887. Nevertheless, his discoveries were known to Henri Deslandres when the latter published his first work on band spectra series in 1886. Deslandres was a junior colleague of Cornu's at the École Polytechnique, and Cornu helped and encouraged him in his work.

In this Deslandres had the use of a Rowland concave grating which had recently been presented to the École Polytechnique by Rowland himself.[71] As the use of Rowland gratings gave rise to a new era in spectroscopic measurement a brief mention must be made of them before considering Deslandres's work.

Henry A. Rowland was the first Professor of Physics at Johns Hopkins University. Like many others he was well aware of the elementary fact that the accuracy of a grating is directly dependent upon the accuracy of the screw of the ruling engine. So Rowland tried, as no doubt his predecessors had tried, to make as perfect a screw as possible— with the result that he constructed a dividing engine which ruled gratings "much more" accurately than any previous engine.[72]

In the ruling of a grating a fine diamond point was drawn across a polished glass, or speculum metal, surface in one direction only. Between each stroke the ruling carriage carrying the diamond point was moved, by means of the screw, through a determined distance in a direction at right angles to that of the stroke. The engine was mechanically driven and once started was not stopped until the ruling was completed. This could take up to six days, depending upon the size of the grating, and during the entire operation every effort was made to keep the temperature constant.

In 1881 the idea occurred to Rowland of ruling gratings on a spherical mirror of speculum metal and he subsequently made a thorough study of the properties of such concave gratings. With plane gratings it is necessary to employ lenses to focus the diffracted rays, and one of the advantages of Rowland's grating is that it renders the use of lenses unnecessary. It was for this and other reasons that one practicing spectro-

[71] For mention of Rowland's gift see A. Cornu, "On the distinction between spectral lines of solar and terrestrial origin," *Phil. Mag.*, 22 (1886): 458–63.
[72] For this information on Rowland's grating I have depended upon E. C. C. Baly, *Spectroscopy* (3d ed.; London, 1929), 1: 27 ff. For the mounting and use of the grating see chapter 7 of the same volume.

scopist would write in 1927 that Rowland's grating "has proved to be one of the greatest inventions ever made in spectroscopy."[73] From the mid-1880s it became part of the standard equipment in the determination of wavelengths.

Deslandres took full advantage of its powerful dispersion in studying the band spectrum given by the light surrounding the negative pole in a nitrogen discharge tube. For one particular band he finds the simple law that *"the interval between successive lines, calculated in vibration numbers or inverse wavelengths are nearly in arithmetic progression."* And he continues that

> An analogous relation has already been pointed out by Piazzi Smith and Herschel . . . but it has been presented by them as an isolated fact. Now this simple law seems to be general and I have verified it, nearly in the same manner, with some secondary deviations, for all the bands which I have been able to resolve into fine rays. The structure is only much less simple in general; for most often the bands are formed not by a single series, but by several equal arithmetical series superposed and confused among one another. The number of series is the same for bands of the same origin.[74]

In a band consisting of more than one overlapping series the vibration numbers of the lines of one series can be obtained by adding a constant to those of the lines of another series. Following H. Kayser and C. Runge I shall refer to this as Deslandres's first law.

The earlier law, pointed out by Deslandres, relating the lines of a given series is called by Kayser and Runge Deslandres's second law. Deslandres was soon to show that it could be expressed by the formula $1/\lambda = Am^2 + a$ where A and a are constants and m takes the successive values 1, 2, 3, etc.[75]

In 1887 Deslandres announced that this second law applied not only to the series of lines within bands but also to the series of bands composing a spectrum. Representing the bands by the lines terminating them on the side of the smallest intervals, and converting the wavelengths to wave numbers, the series of bands is therefore represented by

$$\frac{1}{\lambda} = Bn^2 + \beta.$$

Combining the two expressions we find that the distribution of lines in a band spectrum is given by the expression

---

[73] *Ibid.*, p. 28.

[74] H. Deslandres, "Spectre du pôle négatif de l'azote. Loi générale de répartition des raies dans les spectres des bandes," *Comptes Rendus*, 103 (1886): 375–79; p. 376.

[75] H. Deslandres, "Loi de répartition des raies et des bandes, commune à plusieurs spectres des bandes," *Comptes Rendus*, 104 (1887): 972–76.

$$\frac{1}{\lambda} = Am^2 + Bn^2 + \gamma.^{76}$$

This is Deslandres's third law. It would appear that all three laws were found quite independently of Balmer's work.

In regard to the criticism of Deslandres's work by Kayser and Runge, which we are about to consider, it is interesting to note that Rowland independently discovered the Deslandres–Herschel law in 1885 when he began to map the solar spectrum. In testing the law Rowland first investigated the solar bands A, B, and a, and afterward all of the carbon bands. He found, however, that the law was not as exact as his measurements required, and so he never published it.[77]

It was in 1889 that Kayser and Runge published the results of their investigation of Deslandres's work.[78] They too had employed one of Rowland's "splendid" concave gratings in their determinations. The first Deslandres law they find to be false. The second is only approximately correct, much better agreement being given by

$$\frac{1}{\gamma} = A + Bm^2 + Cm^3 + Dm^4 + Em^5$$

and still better agreement by

$$\frac{1}{\lambda} = A + Be^{cm}\sin(Dm^2).$$

Yet in spite of the "remarkable agreement" of this latter expression with observation, Kayser and Runge believe that the true expression has not yet been discovered. Perhaps, they say, it will be obtained only from theoretical considerations. However, as we shall see in another chapter, there was little prospect of that occurring soon. Finally, Kayser and Runge also find Deslandres's third law to be only approximately correct, but this time they do not offer a more accurate expression.

These were the important advances made in understanding the distribution of lines in band spectra during the period down to 1897. Deslandres subsequently became involved in astrophysical research, while Kayser and Runge, as we are about to consider, gave their energies to the more interesting study of line spectra.

[76] *Ibid.,* p. 973.
[77] Rowland wrote to Deslandres about this—see *ibid.,* p. 972. The details, however, are found in J. S. Ames, "On the relations between the lines of various spectra, with special reference to those of cadmium and zinc, and a re-determination of their wave-lengths," *Phil. Mag.,* 30 (1890): 33–48; p. 40.
[78] H. Kayser and C. Runge, *Über die Spectren der Elemente. Zweiter Abschnitt* (Berlin, 1889), pp. 12–16.

7. The Discovery of Further Line Spectra Formulae: The Work of Rydberg and Kayser and Runge

Prior to taking up the position of professor of physics at the Technische Hochschule in Hannover in the autumn of 1885, Heinrich Kayser had been an assistant to Helmholtz in Berlin. Kayser's interest in spectroscopy was first aroused by a series of three lectures given by another assistant, E. Hagen, in 1881. "Kayser recognized immediately that the study of the spectra of different atoms and molecules would eventually lead to information about atomic and molecular structure and therefore he decided forthwith to turn all his energy to this field."[79] Throughout the following year Kayser gave most of his free time to spectroscopy and as a result he published in 1883 his *Lehrbuch der Spectralanalyse*. It was to be some time, however, before he published any original work. Yet the direction of his interests can be seen from a report on the work of Balmer and Cornu which he gave to the Physical Society of Berlin shortly before leaving that city for Hannover. In Kayser's opinion Cornu's investigation gives an "increased significance" to Balmer's work.[80]

In 1886 Carl Runge also came to Hannover from Berlin, as Professor of Mathematics, and in the following year he and Kayser began a joint search for further spectral series formulae. Prior to this Runge had published several mathematical papers, but would appear to have had no active interest in spectroscopy.

No doubt influenced by Cornu's results, Kayser and Runge began by trying to apply Balmer's hydrogen formula to the many series which exist in the spectra of other elements. But in this they met with no success. However, following a year's study Runge announced at the meeting of the British Association for the Advancement of Science, in 1888, that they had found formulae for groups of lines described by Liveing and Dewar as "harmonic series of lines."[81]

These formulae are given in the report without any explanation of how they were obtained. They are:

$$\lambda = \frac{1}{a + bm^{-1} + cm^{-2}}$$

and

$$\lambda = \frac{1}{a + bm^{-2} + cm^{-4}},$$

[79] See the short biography of Kayser in *Biographical Memoirs of Fellows of the Royal Society*, 1 (1955): 135–40.
[80] See report of June 12, 1885, meeting of Berlin Physical Society in *Nature*, 32 (1885): 312.
[81] C. Runge, "On the harmonic series of lines in the spectra of the elements," *Rep. British Assn.*, 56 (1888): 576–77.

where a, b, and c are constants for the series, and m takes successive integral values. As an example, Runge shows how the eight lines of a lithium series observed by Liveing and Dewar can be represented by the formula

$$\lambda = \frac{1}{4341.4 + 136.5m^{-1} - 11635m^{-2}}$$

where m takes the successive integral values from 3 to 10. In this case the agreement with observation is excellent, but it is not so with the formulae for other series. Regarding the constants in the formula Runge says that possibly they will be found to have some connection with atomic weight.

The same year saw the first of Kayser and Runge's several joint papers on the spectra of the elements.[82] Here they claim to have discovered a formula which represents the spectral series of a number of elements. The formula is not given but Balmer's law for hydrogen is said to be a special case of it. Wavelengths calculated from the formula are in agreement with observed wavelengths, so long as the latter are accurately determined—which "unfortunately" is so in only a small percentage of cases. Errors of three Ångstrom units are to be found even in the measurements of "the dependable observers, Liveing and Dewar." Consequently, Kayser and Runge "resolved to determine anew the spectra of the elements from one end to the other." In undertaking this laborious and difficult task they drew encouragement from the fact that a Rowland grating would permit a hitherto unattainable accuracy.

Kayser and Runge proceeded by mapping a normal spectrum upon which they could base the determinations of all the other spectra. Instead of the customary solar spectrum they selected that of iron. The latter has a suitably wide range of wavelengths, and, furthermore, is always available, whereas the solar spectrum is not. Having completed the normal spectrum they carried out their examination of Deslandres's work, mentioned above, before continuing with line spectra in the group of alkali metals. During the course of this latter work they must have been greatly surprised by the unexpected news that a Swedish mathematical-physicist, Janne Rydberg, had for some time been working along similar lines. Rydberg published an abstract of his results in the *Philosophical Magazine* and *Comptes Rendus* early in 1890.[83] However, the full account of his work appeared only after Kayser and Runge had published the results of their investigations of the alkali spectra. I shall consider the latter first and then Rydberg's work.

82 H. Kayser and C. Runge, *Über die Spectren der Elemente. Erster Abschnitt* (Berlin, 1890).
83 J. R. Rydberg, "On the structure of the line spectra of the chemical elements," *Phil. Mag.*, 29 (1890): 331–37; and *Comptes Rendus*, 110 (1890): 394–97.

Kayser and Runge's method of arriving at further series formulae is to seek by trial and error suitably generalized forms of Balmer's formula.[84] This latter expression,

$$\lambda = \lambda_0 \frac{m^2}{m^2 - 4}$$

may be written in a general form as

$$\frac{1}{\lambda} = A - Bm^2,$$

where A and B are constants. Now, say Kayser and Runge, accepting $1/\lambda$ to be a function of m^{-1} or m^{-2} "it suggests itself" for other elements to expand this formula in a series of only a few terms to obtain an expression which is convergent for increasing m. Convergent, because this is the actual nature of spectral series. "Accordingly one takes expressions of the form

$$\frac{1}{\lambda} = A + Bm^{-1} + Cm^{-2},$$

$$\frac{1}{\lambda} = A + Bm^{-2} + Cm^{-4},$$

or other expressions which, for example, may contain also the first and third or second and third powers. Since the second power occurs in Balmer's formula it suggests itself to take in the general formula the second and fourth powers, and we have in fact found that in general this formula gives the best results."[85]

Kayser and Runge consider the spectra of the alkalis in the order of increasing atomic weight and so begin with lithium, in connection with which Runge had first announced their formulae. The lithium spectrum is now seen to be composed of three series readily distinguished by their appearances. The same is true for all the alkali spectra, even though in each there are lines belonging to none of the series, which nevertheless do not seem to form a series by themselves. Very much later, beyond our period of interest, these lines would be recognized as belonging to a series called the fundamental.[86] Of the three series recognized in 1890 one is composed of distinct and easily reversed lines extending throughout the length of the spectrum. This was named the principal series both by Kayser and Runge and by Rydberg. Of the two remaining series one is composed of lines which are broader and

[84] H. Kayser and C. Runge, Über die Spectren der Elemente. Dritter Abschnitt (Berlin, 1890), pp. 31–33; see also Kayser, Handbuch, 2: 514–15.
[85] H. Kayser, ibid.
[86] By A. Bergmann—see E. Whittaker, A History of the Theories of Aether and Electricity (New York, 1960), 1: 379.

brighter than those of the other. Kayser and Runge name it the first subordinate series and the other the second subordinate series. Rydberg, on the other hand, follows the nomenclature of Liveing and Dewar and from their appearance names these respectively the diffuse and sharp series. As the lines of the alkali spectra are doublets, six single-line series may be distinguished in each. (In 1890 the lithium lines were regarded as being single.)

The wavelengths of the lines of any series having been determined, the next step is to take any three of them together with the corresponding values of m—the selection of which will initially be a matter of trial and error—and to substitute these in one or other of the formulae. One then has three linear equations in A, B, and C and it is a simple matter to solve for these three constants. One next calculates the wavelengths of the remaining members of the series and compares them with the observed values. Should the agreement be poor, one has then either to try the same formula again while ascribing different values of m to the lines, or one can try a different formula. In this nontheoretical and quite mechanical way Kayser and Runge found that the formula

$$\frac{1}{\lambda} = A + Bm^{-2} + Cm^{-4},$$

where the smallest value of the consecutive integers m is 3, best fitted the various series of the alkali spectra. We recall that when Runge first announced their work he cited the formula

$$\frac{1}{\lambda} = A + Bm^{-1} + Cm^{-2}$$

as giving excellent agreement for a lithium series. Two industrious years of observation and calculation had brought a change.

Kayser and Runge mention that it may be objected, as indeed Rydberg will object, that theirs is merely an interpolation formula and that therefore countless other formulae would yield the same results. They claim, however, that this is not the case as they have tried unsuccessfully to represent the principal series of lithium, for example, by means of the formulae

$$\frac{1}{\lambda} = a + bm^{-1} + cm^{-2},$$

and

$$\frac{1}{\lambda} = a + bm^{-1} + cm^{-2} + dm^{-3}.[87]$$

[87] H. Kayser and C. Runge, "Über die Spectren der Alkalien," Ann. der Physik, 41 (1890): 302–20; p. 308.

Yet in a second account of their work they do concede that each arbitrary formula with three constants can be brought to "some" agreement with the observed wavelengths. However, the "question is only with what accuracy one wants to be content."[88] And they claim that their formula is considerably more accurate than any other with the same or a greater number of constants. Certainly it was quite remarkable that the same general formula could be applied to all of the series considered.

This, however, was an achievement which was surpassed by Rydberg in his extensive first paper.[89] Rydberg was a lecturer in both physics and mathematics at the University of Lund. In physics he was interested in theoretical-mathematical questions so he is perhaps more akin to Runge than to Kayser. But Rydberg is perhaps most like Balmer, for while he installed a grating spectrograph of the highest quality he seems always to have relied on the observations of others, including those of Kayser and Runge, in his mathematical investigations. It is said that Rydberg had a "rare" mathematical mind and comparable memory, and that he often calculated with logarithms without using logarithmic tables.[90] For his doctoral degree, obtained in 1879, he had written on the construction of conic sections and later he published a work on the integrals of algebraic equations.

As we shall see, Rydberg commenced his investigations sometime before Balmer published his law in 1885. This was also well before the beginning of the collaboration of Kayser and Runge, whose publications seem to have induced Rydberg to publish his results in what he considered to be an unfinished form.

Rydberg began by making a graphic representation of the relations of the doublets or triplets in each of the spectra of sodium, potassium, magnesium, calcium, and zinc, in the same way that Stoney had represented the hydrogen lines discovered by Huggins. That is, he plotted the wave numbers of the lines of each doublet and triplet against the integers $1, 2, \ldots m$, denoting the position of a line in the spectrum, i.e., first, second . . . mth line. And he found that the lines were represented by curves lying parallel to one another. To consider one of these curves —that representing the series of least refrangible lines of all sodium doublets—Rydberg found that its step-like character could be replaced by two smooth parallel curves when, instead of joining points with

88 H. Kayser and C. Runge, *Über die Spectren der Elemente. Dritter Abschnitt,* p. 32.
89 J. Rydberg, "Recherches sur la constitution des spectres d'émission des élements chimiques."
90 For further biographical details see Manne Siegbahn, "Janne Rydberg, 1854–1919," in *Swedish Men of Science* (Stockholm, 1952), pp. 214–18.

consecutive values of m, one joined all the points with odd and all the points with even values of m. Furthermore, in excluding the third least refrangible doublet from either series one could ascribe the same value of m to corresponding lines of the two series and find the following two series:

m	1	2	3	4	5	6
λ_1	8199	5687	4983	4667	4496	4393
λ_2	6160	5155	4751	4543	4423	—

When now one takes account of the physical character of the lines in each series their significance increases. "In fact Liveing and Dewar, who have observed all the lines cited above, describe those of the first line as diffuse, those of the second as sharp, such that it would have been possible to distinguish the two series independently of the [graphical representation]."[91] And as only the first line of each doublet has been plotted it follows that each of the above series has a corresponding series for the second line of the doublet, giving in all four series for sodium. Rydberg will later also mention the two principal series. He tells us that he discovered independently of Hartley that the separation of doublet or triplet components is a constant in a given series.[92]

Following this considerable simplification, Rydberg tried to give mathematical expression to the form of the series which, from the curves, appeared to be similar for all series. It seemed to him that the curves approached each of the axes asymptotically and that therefore he ought to examine as the most simple form of the function the equilateral hyperbola

$$(\lambda - \lambda_0)(m + \mu) = C,$$

where $\lambda - \lambda_0 = 0$ and $m + \mu = 0$ are the asymptotes, and c is some constant. However, when tested this equation gave only a "very superficial idea of the constitution of spectra." Consequently, Rydberg tried to find other more satisfactory functions. To this end the most significant change in his method of research was to make use of oscillation numbers instead of wavelengths. The oscillation, or wave, number n of a line of wavelength λ is the number of wavelengths contained in unit length. Rydberg's original reason for this change was that it halved the calculations involved with doublet series. It was for a similar reason that Stoney had introduced inverse wavelengths, which differ from wave numbers by only a constant factor.

[91] J. R. Rydberg, "Recherches sur la constitution des spectres d'émission des éléments chimiques," p. 34.
[92] *Ibid.*, p. 11.

Rydberg had from the beginning also studied corresponding separations of consecutive series lines in the hope of finding a regular variation with increasing values of m. However, it was only after he had introduced wave numbers that he discovered a regularity enabling him to find the corresponding lines of different elements. As an example he compares the separations Δn of corresponding lines in the various lithium and sodium series.

		Sharp series		Diffuse series		
	Lithium	Sodium		Lithium	Sodium	
		Series 1	Series 2		Series 1	Series 2
$\Delta_1 n$	3290.2	3165.6	3159.0	5338.3	5386.7	5386.8
$\Delta_2 n$	1694.5	1647.0	1655.0	2476.7	2484.9	2477.0
$\Delta_3 n$				1349.5	1357.9	1363.9

The similarity is even more striking in another family of elements—that of magnesium, zinc, cadmium, and mercury—where again the members of the series of differences Δn decrease rapidly with increasing m. Rydberg postulates the general rule that the differences Δn for pairs of corresponding lines are almost of the same value for the spectra of the elements of the one natural family. Almost, but not quite, and Rydberg finds that the series $\Delta_1 n, \Delta_2 n, \Delta_3 n, \ldots$ for several different elements may be arranged in order such that the terms Δn of one series are all greater than the corresponding terms of the next series below it, and all smaller than those of the next series above it. If one then plots these various series of Δn, with Δn as ordinate and the position-number m of the terms as abscissae, one obtains a series of continuous curves no two of which intersect.

Rydberg then suggests that all of the curves can be made to coincide by means of an appropriate translation of each parallel to the m-axis. For, as far as can be judged, the difference of m for any two curves is a constant for all values of Δn. "If our hypothesis, contrary to all expectations, is not rigorously true, we at least will have a very satisfactory approximation."

Supposing then that the curves are all of the same form and that they differ only in their positions, if the equation of any one of them is $\Delta n = F(m)$ the others will have the form $\Delta n = F(m + \mu)$ where μ is a constant having different values for the several series. So the general expression for the series is $\Delta n = F(m + \mu)$. "However, it must not be forgotten that our conclusions, being drawn from the examination of properties of series, are subject to the same uncertainty as the measurements of the lines of which the series are formed."[93]

[93] *Ibid.*, p. 38.

Returning now to the original series we know that

$$n_m = n_{m+1} - \Delta n;$$

therefore $\qquad n_m = n_{m+1} - F(m + \mu),$

and continuing the sequence we have

$$n_{m+1} = n_{m+2} - F(m + 1 + \mu)$$
$$n_{m+2} = n_{m+3} - F(m + 2 + \mu)$$

$$\underline{\quad} \quad \underline{\quad} \quad \underline{\quad}$$

$$\underline{\quad} \quad \underline{\quad} \quad \underline{\quad}.$$

Now we already know that for $m = \infty$ the value of n always approaches a finite limit which we denote by n_0. Furthermore, the form of the series of Δn, as also that of the corresponding curves, suggests that the limit of Δn, or of $F(m + \mu)$, is equal to zero for $m = \infty$. Therefore in adding together all of the preceding equations we obtain

$$n_m = n_0 - \sum_m^\infty F(m + \mu),$$

where the summation is always a finite quantity. Replacing the summation by $f(m + \mu)$ we have

$$n = n_0 - f(m + \mu)$$

as the general formula for all series. For each particular series n_0 and μ are characteristic constants, but all other constants involved are the same in all the series.

Now the form of the curves obtained by plotting any series of wave numbers n against position numbers m indicate that they probably have two asymptotes, one parallel to each of the n and m axes. The first is given by the line $n - n_0 = 0$, and for $m = \infty$ we have $n = n_0$, or $f(m + \mu) = 0$. The second asymptote must be of the form $m + \mu + c = 0$, where c is a constant having the same value in all series. But as nothing has been determined about the absolute magnitude of μ this asymptote may be represented by $m + \mu = 0$. From which we have $f(m + \mu) = \infty$ for $m + \mu = 0$. The most simple form of $f(m + \mu)$ satisfying these conditions is

$$f(m + \mu) = \frac{C_0}{m + \mu},$$

where, and this is most important, C_0 ought to be a constant common to all series.

However, the equation which is thereby obtained for n, namely,

$$n = n_0 - \frac{C_0}{(m + \mu)},$$

is nothing other than the equation arrived at in the first attempt

$$(n - n_0)(m + \mu_1) = C_1.$$

There is the difference though that observation shows C_1 not to have the same value for all series. Nevertheless, the similarity between the two is sufficient to show that the new formula, as the old, is inadequate.

This time, however, Rydberg apparently decided to try the next simplest form of $f(m + \mu)$ and so examined the equation

$$n = n_0 - \frac{N_0}{(m + \mu)^2}. \tag{1}$$

And, as he tells us, "I was just engaged in studying this equation when I saw published Mr. Balmer's calculation of the hydrogen spectrum."[94] Rydberg took the latter's formula,

$$\lambda = \lambda_0 \frac{m^2}{m^2 - 4},$$

and replacing wavelengths by wave numbers obtained:

$$n = n_0 \frac{(m^2 - 4)}{m^2}$$

or

$$n = n_0 - \frac{4n_0}{m^2}.$$

This coincides with equation (1) in substituting $N_0 = 4n_0$ and $\mu = 0$, "which demonstrates that Balmer's formula is a particular case of equation (1)." Then, using Ångstrom's determinations of the hydrogen lines in Balmer's formula, Rydberg calculates the value of the constant N_0, later to be known as Rydberg's constant, appearing in all series.

Rydberg's next task was to demonstrate that his general formula could represent, with appropriate values of n_0 and μ in each case, all the known series of lines. He worked with the spectra of elements of the first, second, and third groups of the periodic table and found that in general the strongest lines forming series could be represented to a good approximation by his equation. Rydberg, however, was not satisfied:

The exceptional accuracy with which our formula represents the observations gives reason to believe that we have found the true functions of series. However it is not so. The computation of all the known series shows us that in several cases there are rather considerable deviations which greatly exceed the errors of observation. This is why I have taken much trouble to find another more exact formula in making use of the results of the preceding researches.[95]

---

[94] *Ibid.*, p. 41.
[95] *Ibid.*, p. 54.

But while Rydberg was to further develop his series formulae, both in his first masterful paper and in later ones, they nevertheless retained essentially their original form. However, this dissatisfaction on the part of Rydberg is of interest in light of the criticism of his work by Kayser and Runge. Following the publication of the abstract of Rydberg's work Kayser and Runge noted that Rydberg's formula may be expanded as a series, the first three terms of which are $A + Bm^{-2} - 2\mu Bm^{-3}$.[96] Thus the Rydberg and the Kayser–Runge formulae agree in the first two terms and, say Kayser and Runge, by means of suitable selection of A, B, and μ Rydberg's formula gives excellent agreement with observation. However, they have not found it better than their own formula. That which interests them most in Rydberg's work, is his statement that the constant B is the same for all series. They have not found this to be the case, although B, presumably in their formula, is not found to vary much. But taking it as a constant there can be a difference between calculation and observation of as much as 23 Ångstrom units in some cases.

In reply Rydberg says that the Kayser–Runge formula is an interpolation formula whose only goal is to represent the given observations with the greatest possible accuracy.[97] On the other hand, his formula tends in the first place to satisfy the conditions of series which are formed by the difference of wave numbers of lines of ordinary series. He continues that the "most perfect agreement" of the Kayser–Runge equation is a result only of the introduction of three independent constants instead of the two in Rydberg's formula (n_0 and μ). "It would be easy to form equations which represent the series with a much greater precision than what has been attained by Kayser and Runge, if one did not fear to increase the number of constants. But it would be of little importance for the knowledge of the constitution of spectra to calculate series by the aid of ordinary interpolation formulae."[98] Rydberg retained his own equation because among those which fulfill the necessary conditions it is the most simple he has been able to find. Although he has not yet found the exact form of the function of $(m + \mu)$, it has nevertheless been of great importance to his researches to have a simple equation. For it was mainly the simplicity of the formula which enabled him to recognize relationships between series, as we shall shortly see. Rydberg has been able to show that for a given spectrum the constant n_0 should be the same for the sharp and diffuse

96 H. Kayser and C. Runge, *Über die Spectren der Elemente. Dritter Abschnitt*, p. 33; and "Über die Spectren der Alkalien," pp. 318–19.
97 J. R. Rydberg, "Recherches sur la constitution des spectres d'émission des éléments chimiques," pp. 133–34.
98 *Ibid.*

series; and he considers it a "proof of the inferiority" of the Kayser–Runge formula, that values of n_0 calculated from it differ much more among themselves than those calculated on his formula.

These criticisms were included in the full account of Rydberg's work and this was now available to Kayser and Runge, who promptly examined the foundations of Rydberg's derivation. These were judged to be insecure and were duly criticized in a paper, the main purpose of which was to show that the Kayser–Runge formula represented with great accuracy the spectral series of the elements of the second Mendelejeff group.[99] Kayser and Runge's main consideration is given to three hypotheses which they find in Rydberg's treatment. The first two of these are: that the wave numbers of a series of lines converge to a fixed limit with increasing m, and that the difference between successive wave numbers may be considered as the value of a single function of $(m + \mu)$. The first they accept as correct as their own formula is based on it. But the second, which "constitutes the essential point of Rydberg's view," they reject.

It is the opinion of Kayser and Runge that Rydberg's assumption—that the curves obtained by plotting Δn against m for the various series can be made to coincide by appropriate displacements along the m axis—is a hypothesis that can be neither strictly proved nor disproved. The reasons they give are that one knows only simple points of each curve and that when the curves are made to overlap, the points of one curve do not coincide with those of another. To the best of my knowledge Rydberg never replied to this or any of the other objections considered here. But it is clear that Kayser and Runge's criticism in this instance is weak, and Rydberg's procedure quite legitimate. The former were on more solid ground, however, when on the basis of their more accurate measurements they showed several of the curves to intersect. This meant that contrary to Rydberg's contention the one function did not apply to every series. However, Rydberg himself had already admitted that he had not yet found the true function, and after all was his goal not the general form of series rather than an exact mathematical representation of each series line?

The third hypothesis discussed by Kayser and Runge is one which they incorrectly see in Rydberg's work. It involves Rydberg's use of the Balmer formula, and the burden of Kayser and Runge's criticism is that Rydberg's function $f(m + \mu)$ must be a complex rather than a simple one.

The lines of battle were drawn and they were to exist well beyond

99 H. Kayser and C. Runge, *Über die Spectren der Elemente. Vierter Abschnitt* (Berlin, 1891), pp. 61–70.

the period of our present interest. In 1900 it was Rydberg rather than Kayser or Runge who, at the International Congress of Physics held in Paris, gave a report on the current state of spectral series researches—a fact which is indicative of the trend of contemporary judgment.[100] In the body of his report Rydberg employs his own formula exclusively, merely including the Kayser–Runge formula in a useful annotated bibliography. On the other hand, Kayser gives first place in the second volume of his *Handbuch der Spectroscopie* (1902) to the Kayser–Runge formula and criticizes Rydberg's work along much the same lines as before.[101]

However, at least by 1897, where our present period of interest ends, a trend had begun in favor of Rydberg's formula. In that year two prominent spectroscopists declared their preference for it. One was Arthur Schuster, who remarked that:

> The greater simplicity in form of $N = A - B/n^2 - C/n^4$ adopted by Kayser and Runge, and the ease with which the constants may be calculated would, independently of other considerations, lead us to give the preference to this form. The advantage is not, however, great and disappears when we have to distinguish cases where we must substitute n^3 for n^4. The reason why I prefer Rydberg's form at present lies in the fact that it adapts itself better to bring out the regularities as well as the irregularities of distribution of lines independently of any special form of equation.[102]

The other was Balmer—in a paper later described by Kayser as "eine recht werthlose Abhandlung."[103]

According to Balmer the Kayser–Runge formula cannot be regarded as the true expression for the distribution of spectral lines.[104] Balmer maintains that the latter has to be a closed or finite function whose denominator can be expanded in an infinite series. Balmer himself has arrived at the expression

$$n = a \, \frac{(m + c)^2}{(m + c)^2 - b} \quad \text{or} \quad \tau n = A - \frac{B}{(n + c)^2},$$

and he points out its similarity to Rydberg's formula. The only difference is that Balmer's constant B has a different value for each element.

[100] J. Rydberg, "La distribution des raies spectrales," *Rapports presentés au Congrès International de Physique reuni à Paris en 1900* (Paris, 1900), 2: 200–24.

[101] H. Kayser, *Handbuch,* 2: 509–59.

[102] A. Schuster, "Professor C. Runge and F. Paschen's researches on the spectra of oxygen, sulphur and selenium," *Nature,* 57 (1897–98): 320–21.

[103] H. Kayser, *Handbuch,* 2: 567.

[104] J. Balmer, "A new formula for the wave-lengths of spectral lines," *Astrophys. J.,* 5 (1897): 199–209.

Balmer concludes by indicating his belief that Rydberg may have found the true closed function.

Rydberg's own doubts about this being so must have been greatly mitigated by the relationships which he found to exist between the different series of the one spectrum. To consider these we must return to Rydberg's first paper.

We recall that three types of series had been distinguished—principal, sharp, and diffuse. Considering now the more general case of triplet series, Rydberg shows that the three overlapping sharp series are given by the equations:

$$\text{Series 1: } n = n_1 - \frac{N_0}{(m + \mu)^2}$$
$$\text{Series 2: } n = n_2 - \frac{N_0}{(m + \mu)^2}$$
$$\text{Series 3: } n = n_3 - \frac{N_0}{(m + \mu)^2}.$$

Each series has a different limit, but N_0 and μ are common to all three and m takes the same value for corresponding lines. The first series is the least refrangible one, and all three are referred to as the sharp group. The diffuse series are given by these same equations, except that μ now has a different value. As mentioned earlier, Rydberg had found corresponding sharp and diffuse series to have the same limit n_0.

Rydberg further finds—although only in the case of the sodium, potassium, and rubidium doublet spectra—that the two principal series of the one element have the same value of the constant n_0. Thus the two series are given by:

$$n = n_0 - \frac{N_0}{(m + \mu_1)^2},$$

and

$$n = n_0 - \frac{N_0}{(m + \mu_2)^2}.$$

Finally, Rydberg notices that the difference in wave number between the two components of the most refrangible doublet in a principal series has the same value as the constant separation of doublet components in sharp and diffuse series. Thus we have from above

$$n_2 - n_1 = \left(n_0 - \frac{N_0}{(1 + \mu_1)^2}\right) - \left(n_0 - \frac{N_0}{(1 + \mu_2)^2}\right) = \frac{N_0}{(1 + \mu_2)^2} - \frac{N_0}{(1 + \mu_1)^2}.$$

"It seems however that there is a much more intimate relation between the two species of series in question: without doubt they are only *different parts of the one series of two variables*. I hope at least to be able to give weighty reasons for the existence of this relation although it is not possible for one at present to demonstrate it with absolute certainty."[105] Without going into detail I shall merely say that Rydberg shows that the sharp and diffuse series of lithium, sodium, and potassium may be written approximately as:

Sharp series:
$$n = \frac{N_0}{(1 + \mu_{12})^2} - \frac{N_0}{(n + \sigma)^2}$$

Principal series:
$$n = \frac{N_0}{(1 + \sigma)^2} - \frac{N_0}{(n + \mu_{12})}.$$

The progress subsequently to be made in this direction was fundamental to the development of the quantum theory of spectra. And it was in this way that the superiority of Rydberg's approach was conclusively established. To this end one other important step was taken by 1897—this was the discovery of the relationship later to be known as the Rydberg–Schuster law. It was first announced by Rydberg in 1896 and again in the following year independently by Schuster, who had apparently been working with the Kayser–Runge formula.[106] The relationship is that the difference between the common limit of the diffuse and sharp series and the limit of the corresponding principal series, measured in wave numbers, gives the common first term of the principal and sharp series. Schuster considered this "the most important fact yet brought to light concerning molecular vibrations."

It is not at all obvious that this relationship is included in the relationship between the principal and sharp series suggested by Rydberg in 1890 and given above. But we see from the latter that the difference of the limits of the sharp and principal series is

$$\frac{N_0}{(1 + \mu_{12})^2} - \frac{N_0}{(1 + \sigma)^2}, \tag{A}$$

or
$$\frac{N_0}{(1 + \sigma)^2} - \frac{N_0}{(1 + \mu_{12})^2}. \tag{B}$$

And further that for $m = 1$ (A) corresponds to the first line of the sharp series and (B) to the first line of the principal series. Certainly this circumstance was not obvious to Rydberg either, for six years were to elapse before he recognized it. However, once he had done so Rydberg implied that he had known it all along and remarked that it therefore

[105] J. R. Rydberg, "Recherches sur la constitution des spectres d'émission des éléments chimiques," p. 60.
[106] J. Rydberg, "The new elements of clevite gas," *Astrophys. J.*, 4 (1896): 91–96; A. Schuster, "On a new law connecting the periods of molecular vibrations," *Nature*, 55 (1896–97): 200–1.

seemed "impossible for Professor Schuster to have his claim to a second discovery on the law not having been sufficiently published."[107]

These relationships between the various series of the one spectrum were used by Kayser and Rydberg to identify the famous series of lines discovered in the spectrum of ζ Puppis in 1896 by the Harvard astronomer, Edward C. Pickering.[108] At first Pickering found that the six lines could be represented approximately by a modified Balmer formula:

$$\lambda = 4650 \, \frac{m^2}{m^2 - 4} - 1032,$$

for $m = 10, 9, 8, 7, 6$, and 5. He consequently regarded the series as belonging to a hitherto unknown element. However, Pickering soon found that, using Rowland's standard wavelength determinations and replacing m by $m/2$ in Balmer's formula, the resulting formula

$$\lambda = 3646.1 \, \frac{m^2}{m^2 - 16}$$

could represent both the hydrogen and the new series—the former being given when the even integers 6, 8, 10, 12, etc., are substituted for m, and the latter when the odd integers 5, 7, 9, 11, etc., are substituted.[109] Thus the Pickering series now appeared to belong to hydrogen. This view had already been adopted by Kayser, who argued, however, that the representation of the two series as a single series was incorrect. To him they were the sharp and diffuse series of hydrogen.[110] This latter was also the opinion of Rydberg, who succeeded in showing that the two series could be represented by different formulae having a common limit n_0—in other words that they were in fact the sharp and diffuse series of hydrogen.[111] The view that Pickering's series was the diffuse hydrogen series persisted until 1913, when Bohr demonstrated that it really belonged to ionized helium.[112] Nevertheless, Rydberg scored a triumph for his work by calculating the principal series of hydrogen from the sharp series, and identifying the first line of the former in a stellar spectrum.[113]

---

[107] J. Rydberg, "The new series in the spectrum of hydrogen," *Astrophys. J.*, 6 (1897): 233–38; p. 236 ff.

[108] E. C. Pickering, "Stars having peculiar spectra. New variable stars in Crux and Cygnus," *Astrophys. J.*, 4 (1896): 369–70.

[109] E. C. Pickering, "The spectrum of ζ Puppis," *Astrophys. J.*, 5 (1897): 92–94.

[110] H. Kayser, "On the spectrum of ζ Puppis," *Astrophys. J.*, 5 (1897): 95–96; and "On the spectrum of hydrogen," *Astrophys. J.*, 5 (1897): 243.

[111] J. Rydberg, "The new series in the spectrum of hydrogen," pp. 236 ff.

[112] E. Whittaker, *A History of the Theories of Aether and Electricity* (New York, 1960), 2: 114.

[113] J. Rydberg, "The new series in the spectrum of hydrogen," pp. 236–37.

The spectral series field was dominated throughout the 1890s by Rydberg, Kayser, and Runge, and their common goal was to resolve the complex line spectra of all the known elements into series of lines. The remarkable similarity of the series found with the elements of lower periodic grouping and lower atomic weight led them to seek similar series among the more complex spectra of elements of higher periodic grouping and higher atomic weight. But as early as 1893 Kayser and Runge found that the lines of certain elements of the fourth and fifth periodic groups were ordered in quite a different way.[114] Here there were no series of doublets and triplets arranged in a series converging toward a limit. Instead, the lines could be ordered in rows and columns, such that there is a constant difference of wave number between the corresponding members of any two rows or columns. In 1897 Kayser found similar series in the spectra of elements of the eighth periodic group.[115] In the same year Rydberg also found such series in the addition to the eighth group, argon, as well as in a member of the first group, copper.[116] But even by 1900 the mathematical expression of the law, or laws, involved had still not been found.[117] The state of knowledge of these spectra in 1897 may be compared to that of the alkali spectra in the 1880s, when Rydberg and then Kayser and Runge began their investigations.

Their domination of the field must be partly explained by the circumstance that, according to Schuster, relatively few physicists were attracted to the subject.[118] As Kayser and Runge acknowledged, the work involved was "unusually laborious and difficult," and we might also add somewhat dull. A multitude of precise observations, or mechanical computations, or both, were required. However, in asking why then Rydberg, Kayser, Runge, Deslandres, and others undertook such investigations, we find some interesting answers.

We shall begin with Rydberg, who affords us a unique view of the physical-chemical scene about 1890. He says that among contemporary

114 H. Kayser and C. Runge, *Über die Spectren der Elemente. Siebenter Abschnitt* (Berlin, 1893).
115 H. Kayser, "Über die Bogenspectren der Elemente der Platingruppe," *Abhandl. Berlin Akad.* (1897).
116 J. Rydberg, "On the constitution of the red spectrum of argon," *Astrophys. J.*, 6 (1897): 338–48; and "On triplets with constant differences in the line spectrum of copper," *Astrophys. J.*, 6 (1897): 239–43.
117 J. Rydberg, "La distribution des raies spectrales," pp. 220–21.
118 A. Schuster, "On a new law connecting the periods of molecular vibrations," p. 200. Two who gave unsuccessful and unproductive formulae were: Adolph Erik Nordenskiöld, "Sur un rapport simple entre les longuers d'onde des spectres," *Comptes Rendus,* 105 (1887): 988–95; and Joseph S. Ames, "On the relations between the lines of various spectra," p. 47.

scientists the hypothesis is generally adopted that all physical phenomena are uniquely due to the motions of matter—"which leads one to believe that physics could one day become, the same as chemistry, a mechanics of atoms."[119] However, there is a very serious obstacle to the realization of this idea, namely, that very little is known of the constitution of these atoms—due no doubt to the fact that the subject has not yet been sufficiently studied. Rydberg feels that he does not exaggerate in saying that "a systematic and comprehensive study of the properties of matter, with the object of realizing the constitution of elements and of developing from it a mechanism of atoms, has not yet been started."[120] But this work must be undertaken if science is to progress. Gravitation, chemical affinity and cohesion, heat and light, electricity and magnetism: these form the great isolated parts of physical science of whose relations we have only vague ideas.

If all these phenomena have their origin in different motions of matter it is absurd to believe that we will ever be able to understand them without knowing the constitution of this same matter, the atoms and the ether. That would be as reasonable as a physiology without anatomical difficulties. But in the present case there are particular difficulties; in fact, as we are incapable of recognizing by our senses the last parts of matter, it is necessary for us to deduce from the entire assembly of phenomena at the same time the properties of motion and the constitution of the matter which moves.[121]

Mendelejeff's discovery of the periodic system of elements had provided a new point of departure of "great importance" for the work in question here. During the early and mid-1880s Rydberg had tried to give a start to the necessary research by determining more exactly the periodic relation between the specific weights of elements and their atomic weights.[122] He found that the relationship could approximately be expressed by a series of sines of variable coefficients. Beyond that he concluded that the periodicity of many physical coefficients ought to depend on the force, which acts between two similar or dissimilar atoms, being a periodic function of the atomic weight. In addition it seemed to Rydberg very probable that chemical cohesion, adhesion, and affinity, depended ultimately on the periodic movements of atoms.

The most natural course would therefore be to study the periodic movements in general, and since the spectra of chemical elements are due to movements

[119] J. Rydberg, "Recherches sur la constitution des spectres d'émission des éléments chimiques," p. 8.
[120] Ibid.
[121] Ibid., p. 9.
[122] J. Rydberg, "Om de kemiska grundämnenas periodiska system," Bihang till Konliga Svenska Vetenskaps-Akademiens Handligar, 10 (1884–85), No. 2, 31 pp.

of this type, we find ourselves [in] the domain of spectrum analysis. It is true that we cannot know whether these periodic motions are the ones we seek primitively, but a study of these vibrations will in every instance give us information of great value on the constitution of atoms and will bring us closer to our goal than any investigation of a physical coefficient.[123]

These then were some reasons why Rydberg took up his spectral research, but there were still others. An equally important reason was the superiority of spectral over other physical data—they were "without comparison the richest and most uniform of all relating to all of the known elements." But the most important property of the data is that they "relate to the motions of the least parts of matter, the atoms themselves, in a way that we can expect . . . to find the most simple functions to express the relations between the form of moving bodies, their dimensions and the active forces."[124]

Thus Rydberg's work on series had as its object "a more exact knowledge of the nature of the constitution of atoms."[125] Did he then, before 1897, in any way realize his goal? Initially, the similarity between the simple Balmer formula, expressed in Rydberg's form, and the more complex formula for the series of other spectra, led Rydberg to remark that "this property would confirm without doubt the hypothesis that the elements are compound, and that hydrogen forms the matter of which they are built in the first place."[126] But this idea had to be abandoned with the failure of the hope that all series would be found to have the same form.

Rydberg's optimism about obtaining knowledge of the anatomy of matter in this way was perhaps unwarranted. We recall that Schuster had earlier compared this hope to that of obtaining knowledge of the constitution of a bell merely from the notes it emitted. But Rydberg's optimism was shared by Deslandres who remarked that once the true mathematical function expressing the distribution of a series of lines had been attained, then one was at a point where, with regard to the constitution of matter, mathematical analysis and number theory could be "very useful."[127] However, unlike Rydberg, Deslandres gave only a few years to the study of series before abandoning it for astrophysical research.

As Kayser and Runge devoted their careers to spectroscopic research

123 J. Rydberg, "Recherches sur la constitution des spectres d'émission des éléments chimiques," p. 9.
124 *Ibid.*
125 *Ibid.*, p. 8.
126 *Ibid.*, p. 140.
127 H. Deslandres, "Loi de répartition des raies et des bandes," p. 976.

we would expect their motives to be similar to Rydberg's. And while in general they are—we recall Kayser's motives mentioned at the beginning of this section—there is nevertheless an interesting difference of objective. Like all other spectroscopists, Kayser and Runge accepted that spectral vibrations are produced in the first place by atoms. They believed, however, that their studies would afford knowledge not of the constitution of atoms but of the nature and condition of molecules.[128] The considerations of section 4 of chapter II enable us to understand this difference.

In addition, Kayser and Runge shared the other belief, also mentioned in the previous chapter, that the more complex line spectra are produced not by one but by several molecules. That is, they believed that such spectra are composed of several overlapping spectra. And because of this they had been skeptical when commencing their researches of finding a single formula which would account for all the lines in a given spectrum.[129] The subsequent discovery of six different series in the alkali spectra, for example, seemed to bear out their belief. And they regarded each series as belonging to a different molecule. On the other hand, Rydberg's discovery of relationships between the various series of the one spectrum no doubt encouraged the opposite belief, namely, that the one unit, in Rydberg's case the atom, was capable of producing all of the series.

---

[128] H. Kayser and C. Runge, *Über die Spectren der Elemente. Erster Abschnitt*, pp. 4–5.
[129] *Ibid.*, p. 7.

IV

||||||||||||||||||||||||||||||||||

Evolution of Qualitative Spectral Theories

THE CONSIDERATIONS of the previous chapter, and to a lesser degree those of chapter II, raise the question of a mathematical theory of spectra. However, before the attempts to give such a theory can be properly considered account must first be taken of other discoveries, besides spectroscopic ones, which during the last quarter of the century, either independently or in conjunction with spectroscopic considerations, led the physicist to change his conceptions of atoms and molecules. In this respect the kinetic theory of gases and, later, the electromagnetic theory of light were most important. This chapter discusses the evolution of qualitative theories of spectra brought about by discoveries successively made in these areas. The question of a mathematical theory of spectra is left for consideration in the following chapter.

1. The Atomic-Molecular Theory of Spectra in Britain

As was the case in discussing the development of the original molecular theory of spectra, so here in considering additional evidence for the atomic-molecular theory, an examination of the further evolution of Maxwell's thought is most illuminating.

In the spring of 1875, and some months before Salet first proposed the atomic-molecular theory of spectra, Maxwell spoke before the Chemical Society of London on the physical, or dynamical, evidence for the molecular constitution of bodies.[1] Having given the evidence support-

---

[1] J. C. Maxwell, "On the dynamical evidence of the molecular constitution of bodies," *Nature*, 11 (1875): 357–59, 374–77; in *Scientific Papers of J. C. Maxwell*, 2: 418–38.

ing the molecular theory he then considered what he described as "the greatest difficulty" which this theory had yet encountered. We shall consider this closely because it bears directly upon a profound change in outlook which was to occur before the year was out.

The difficulty involves the old problem of matching the calculated and observed ratios of the specific heats. This, incidentally, was to be a recurrent problem down to the advent of quantum theory—although as we shall see later some progress was made toward its solution before then.

Maxwell treats the problem much more thoroughly than before. Assuming that a molecule is composed of atoms and that each atom is a "material point," then each atom may move in three different and independent ways corresponding to the three dimensions of space. Thus the number of variables required to determine the position and configuration of all the atoms of the molecule is three times the number of atoms. Maxwell assures us, however, that it is not essential to the mathematical investigation to assume that the molecule is made up of atoms. All that need be assumed is that the position and configuration of the molecule can be completely expressed by a certain number of variables. If we suppose this number to be n, then three variables are required to determine the center of mass of the molecule, and the remaining $(n - 3)$ to determine its configuration relative to the center of mass.

Maxwell continues that if the molecule is capable of changing its form under the action of impressed forces then it must be capable of storing up potential energy. And if the forces are such as to ensure the stability of the molecule, the average potential energy will increase when the average energy of internal motion increases. Hence, as the temperature rises the increments of the energy of translation, the energy of internal motion, and the potential energy, will be as 3, $(n - 3)$ and e respectively—where e is some positive quantity depending on the law of force which binds together the constituents of the molecule.

With this Maxwell finds the specific heat at constant volume to be given by

$$\frac{1}{2J} \frac{p_0 v_0}{273} (n + e),$$

and the specific heat at constant pressure by

$$\frac{1}{2J} \frac{p_0 v_0}{273} (n + 2 + e),$$

where J is the mechanical equivalent of heat and p_0 and v_0 the pressure and volume respectively at zero degrees centigrade.

Now, says Maxwell, if the actual values of specific heats as determined by Victor Regnault are considered together with the above expressions, then it is found that $(n + e)$ for air and several other gases cannot be more than 4.9. The same result is obtained on comparing the ratio of the calculated specific heats, $(n + 2 + e) / (n + e)$, with the ratio as determined by experiment, namely, 1.408. For carbonic acid gas and steam, however, the value of $(n + e)$ is greater than 4.9.

"And here we are brought face to face with the greatest difficulty which the molecular theory has yet encountered, namely, the interpretation of the equation $n + e = 4.9$."[2]

Maxwell explains the nature of the difficulty. First, suppose that the molecules are atoms—"mere material points, incapable of rotatory energy or internal motion"—then the values of n and e are respectively three and zero, and the ratio of the specific heats is 1.66, which "is too great for any real gas."[3] This last is a most important point, as we shall shortly see.

Moreover, continues Maxwell, spectroscopy teaches that a molecule can execute vibrations of constant period. "It cannot, therefore, be a mere material point, but a system capable of changing its form."[4] Such a system cannot have less than six variables, that is, must consist of at least two atoms. Now, according to calculation, the greatest possible value of the ratio of specific heats for a gas whose molecules consist of more than one atom is 1.33 (the case when there are two atoms and $n = 6$). But this value is too small for several gases, including hydrogen, oxygen, and nitrogen.

Nevertheless, the difficulty is greater still. Spectroscopy teaches that some molecules execute a great many different vibrations. "They must, therefore, be systems of a very considerable degree of complexity, having far more than six variables." And it is seen from the expression for the ratio of the specific heats, $(n + 2 + e) / (n + e)$, that as n increases this ratio becomes smaller. Yet the calculated ratio is already too small when the molecule is taken to consist of only two atoms. Hence every additional degree of complexity which is attributed to the molecule can only increase the difficulty of reconciling the observed with the calculated ratio of specific heats.

As we noted earlier in chapter II, Boltzmann had previously suggested a way out of this difficulty: namely, that the high values of the specific heats might be explained by taking account of the mutual action between the molecules and the surrounding ether. Maxwell,

2 *Ibid.,* p. 433.
3 *Ibid.*
4 *Ibid.*

however, disagrees—saying that bringing in the ether would have the effect of further increasing the specific heats and so compounding the difficulty.

Maxwell's best remembered statement on the structure of matter was published under the rubric "Atom" in the ninth edition of the *Encyclopaedia Britannica* late in the same year, 1875. On reading through "Atom" we find an account of the original molecular theory of spectra. This does not surprise us, for the atomic-molecular theory had been proposed only that summer and still remained to be accepted. Our interest is then aroused by a long section entitled "On the Theory of Vortex Atoms." This is an atom which Maxwell has never before considered in connection with spectroscopy or the kinetic theory of gases, even though its existence had been postulated by William Thomson some eight years earlier in relation to these latter.

Maxwell describes the mathematical basis and some of the properties of vortex atoms and then makes the startling statement that: "The conditions which must be satisfied by an atom are—permanence in magnitude, capability of internal motion or vibration, and a sufficient amount of possible characteristics to account for the difference between atoms of different kinds."[5] Of these three conditions, Maxwell had hitherto accepted only the first and perhaps also the third, although with regard to the latter we have noted in chapter II the possibility of his previously admitting only one type of ultimate atom. But the second condition is utterly new, and stands in contradiction to the purely molecular theory of spectra which he has espoused only a few pages earlier.

In 1873 Maxwell had declared himself a disciple of Lucretius and was still such when he addressed the London Chemical Society, but now he states that "the small hard body imagined by Lucretius" fails to account for "the vibrations of a molecule as revealed by the spectroscope."[6] Thus the Lucretian atom is to be rejected, and likewise the Boscovichean atom:

The massive centres of force imagined by Boscovich may have more [than the Lucretian atom] to recommend them to the mathematician, who has no scruple in supposing them to be invested with the power of attracting and repelling according to any law of the distance which it may please him to assign. Such centres of force are no doubt in their own nature indivisible, but then they are also, singly, incapable of vibration. To obtain vibrations we must imagine molecules consisting of many such centres, but, in so doing, the possibility of

---

5 J. C. Maxwell, "Atom," in *Encyclopaedia Britannica* (9th ed.; Edinburgh, 1875–89), 2 (1875); in *Scientific Papers of J. C. Maxwell,* 2: 445–84; p. 470.
6 *Ibid.*

these centres being separated altogether is again introduced. Besides, it is in questionable scientific taste, after using atoms so freely to get rid of forces acting at sensible distances, to make the whole function of the atoms an action at insensible distances.[7]

On the other hand, Maxwell finds that Thomson's vortex atom "satisfies more of the conditions than any atom hitherto imagined."[8]

Thus in "Atom" we witness a radical change in Maxwell's thought on atoms and consequently on the origins of spectra. But Maxwell gives us no clue as to the cause of this change. Indeed, he does not even point out to his readers that he has rejected the molecular theory of spectra given earlier. What then was the cause?

Several months before "Atom" was published, the results of a most important experiment in the kinetic theory of gases were given in the *Berichte der Deutschen Chemischen Gesellschaft zu Berlin*.[9] August Kundt and Emil Warburg, respectively ordinary and extraordinary professors of physics at the University of Strassburg, had found the ratio of the specific heats of mercury vapor to have the remarkable value of 1.66. We recall that this was a value which Maxwell had described as "too great for any real gas." It pointed to the fact that the mercury molecule consisted of a single atom! This was quite in keeping with chemical, though not with spectroscopic, thought. On the molecular theory of spectra a single atom could not produce a spectrum, and yet mercury vapor had a beautiful line spectrum. This result, then, was the cause of the radical change in Maxwell's thought.

It might be objected here that as the Kundt–Warburg result was not published in a major journal until 1876, then Maxwell might after all be accepting Salet's atomic-molecular theory of spectra proposed a few months before "Atom" was published. For according to this theory line spectra are produced by atoms. But these atoms are chemical atoms, regarded by most physicists as aggregates of ultimate atoms —the relative motions of the ultimate atoms within the chemical atoms giving rise to spectra. Furthermore, degrees of freedom are not considered in Salet's theory and so constitute no problem. Thus this view did not present the difficulty which the Kundt–Warburg result presented. The latter challenged the idea of certain of the earlier physical molecules, as also chemical atoms, being aggregates of elastically connected ultimate atoms with many degrees of freedom. Rather than

[7] *Ibid.*, p. 471.
[8] *Ibid.*
[9] A. Kundt and E. Warburg, "Ueber die specifische Wärme des Quecksilbergases," 8 (1875): 945.

being the instrument of change, Salet's theory was in part the object of change.

In the remainder of this section I shall consider the Kundt–Warburg experiment and then the history of the mechanical vortex atom theory of matter. The latter, as will be seen, became the accepted theory of matter in Britain and remained so, with fruitful consequences, down to the discovery of the electron. Then in the following section, I shall consider the effect of the Kundt–Warburg result on continental European thought, which, with one important exception, never accepted the vortex atom theory.

Kundt and Warburg were aware of the deep trouble that the kinetic theory found itself in on the matter of specific heats.[10] They were also aware of the existence of gases "which according to chemical reactions are monatomic, if one regards hydrogen as diatomic." Such a gas was mercury vapor. With the "importance of the gas theory in mind" Kundt and Warburg thought that if they could obtain the value 1.66 for the ratio of specific heats of mercury vapor—a value which, as we have seen, theory predicted on the assumption of monatomic molecules—then the fundamental principles of the gas theory would be vindicated and the disagreement found with other gases could be attributed to an incorrect application of these principles.

The principle of their experiment is as follows. The velocity v of sound in a gas is given by the equation

$$v = n\lambda = \sqrt{\frac{e}{d}(1 + at)\frac{C_p}{C_v}}$$

where λ is the wavelength of the note and n its frequency, e the coefficient of isothermal elasticity of the gas, $d/(1 + _a t)$ its density at $t°C$, and C_p and C_v the specific heats at constant pressure and constant volume respectively. Within the limits of Boyle's law e is the same for all gases. If then a source of given pitch is made to produce vibrations in two different gases of density d_1 and d_2 under the same conditions of temperature and pressure, the relationship between the corresponding wavelengths λ_1 and λ_2 is given by

$$\frac{\lambda_1^2 d_1}{\lambda_2^2 d_2} = \frac{C_{p1}/C_{v1}}{C_{p2}/C_{v2}},$$

where C_{p1}, C_{v1}, and C_{p2}, C_{v2} are the corresponding specific heats. Thus knowing the ratio of the specific heats for air to be 1.405 and knowing

[10] A. Kundt and E. Warburg, "Ueber die specifische Wärme des Quecksilbergases," *Ann. der Physik*, 157 (1876): 353–69.

its density, Kundt and Warburg could use these data to find the ratio of the specific heats for mercury vapor.

The apparatus consisted of a long glass tube which was filled with mercury vapor. Through one end of the tube a glass rod was sealed at its mid-point to lie along the axis of the tube. When the part of the rod outside the tube was stroked with a suitably moistened cloth it vibrated longitudinally and set up stationary waves in the gas within the tube. The length of these waves could be determined by dusting the inside of the tube with lycopodium powder which is thrown up into small heaps at the nodes. The distance between successive nodes is half a wavelength.

Kundt and Warburg found the ratio of the specific heats for mercury vapor to be 1.186 times the ratio for air, or 1.666—giving excellent agreement with theory. Thus chemistry and the kinetic theory of gases seemed to agree that the molecules of mercury vapor are monatomic.

For the physicist-spectroscopist, however, the statement that a molecule having three degrees of freedom produces a spectrum was really self-contradictory. Such a molecule has no internal degrees of freedom and consequently cannot produce even a single spectrum line. Curiously, Maxwell overlooked this in replacing the Boscovichean and Lucretian atoms by the vortex atom. He was in the paradoxical position of accepting and at the same time rejecting the Kundt–Warburg result.

The problem was that two facts had to be accommodated. First, kinetic theory taught that mercury vapor was monatomic; and second, mercury vapor gives a line spectrum. The first forced Maxwell to reject the idea of mercury molecules, or chemical atoms, being aggregates of elastically connected Boscovichean or Lucretian atoms, for such aggregates would have many more than three degrees of freedom. He had to accept that mercury molecules are monatomic. But then they cannot be single Boscovichean or Lucretian atoms, for these do not vibrate and yet the second fact demands that they should. They must then be vortex atoms which do vibrate. However, in accommodating this second fact Maxwell has now rejected the first, which does not allow for any internal vibrations.

In "Atom" Maxwell does not say just how a vortex atom produces a spectrum. For that we shall have to turn to William Thomson and his first paper on vortex atoms, which appeared in 1867. But before becoming involved with vortex atoms let us first consider another British reaction to the Kundt–Warburg result.

Writing in the *Philosophical Magazine* in 1875, Robert Bosanquet, a lecturer in mathematics at Oxford, says that the objection to this result is that "the spectroscope tells us that vibrations go on inside the molecule (or atom)." He concludes that all that can be said

> is that the theory cannot at present take count of these intramolecular (intraatomic) vibrations. We have no theory about any of these matters that is quite perfect; but the gradual advance has consisted, and is likely to consist, in the recognition of general correspondences, as in the case of the dynamical theory of optics. From this point of view we may admit the result of Kundt and Warburg as one of interest and importance, while fully keeping in view the incompleteness of the dynamical explanation it affords.[11]

This, in the light of later events, was a very sensible position. Furthermore, Bosanquet now saw his way to solving the other specific heat problems.

It is but a step, he says, from the case of a smooth hard sphere, as he along with Kundt and Warburg pictured the atom, to the case of a smooth hard solid of revolution, which we may conceive of as formed of two small hard spheres rigidly connected, or as a cylinder, or as having many other forms. Such a molecule would have five degrees of freedom—three of translation and two of rotation ("for by the hypothesis that the body is a solid of revolution, perfectly hard and smooth, *vis viva* of rotation about the axis of revolution is excluded"). With five degrees of freedom for a diatomic molecule the calculated ratio of the specific heats gives excellent agreement with the observed value of about 1.4.

Bosanquet realized that this would meet with objections from the accepted molecular theory of spectra which, we recall, taught that the atoms of a physical diatomic molecule are somehow elastically connected and that the molecule has six degrees of freedom. So he endeavored to anticipate and answer such objections.

> If we ask, why must we suppose the two atoms joined by elastic forces and not rigidly? We are told that it is the vibrations of the atoms that do the work we see in the spectroscope lines. But these lines occur in the vapour of mercury, for which our explanation fails to suggest any collocation of atoms more than one in the molecule. Why not, then, admit that the lines are produced by something within the atom which we cannot at present account for, just as we cannot decompose it chemically?
>
> There is nothing to prevent us from forming the conception of very small amounts of energy existing in some way or other within an atom (regarding this as already complex in a manner which we do not understand), though we are unable to give any accurate account of the way in which the movement is originated or maintained.[12]

---

[11] R. Bosanquet, "The velocity of sound, and ratio of specific heats, in air," *Phil. Mag.*, 3 (1877): 271–78.

[12] *Ibid.*, pp. 276–77.

Bosanquet's attitude was typical of that of most chemists—Salet, for example—and, as we shall see, of many continental European physicists, most of whom were never very specific about their atomic models. These individuals were quite unlike Maxwell and certain other physicists, who had definite atomic and molecular models and who also accepted a theory of spectra which had been created in compliance with kinetic theory considerations. The former were either not aware of, or could with equanimity ignore, implications of the Kundt–Warburg result which were all too obvious to the latter. It is nevertheless curious, perhaps, that Bosanquet does not mention the vortex atom, for it was the only physical atom which offered him "something within the atom." But then it would appear that in Britain this latter came into vogue from relative obscurity only after Maxwell's "Atom" was published. Prior to this time only William Thomson and his friend Peter Guthrie Tait had espoused the vortex atom theory.[13]

The source of inspiration of the vortex atom theory of matter was a pioneering mathematical investigation in hydrodynamics published in 1858 by Hermann von Helmholtz.[14] Helmholtz demonstrated in a series of theorems that in a perfect fluid, infinite in extent, closed rings are formed, and that these vortex rings are totally immune to destruction or dissipation, invariable as to strength (a quantity representing the product of the cross section of the ring and its angular velocity about the circular axis), and subject to specified rates of rotational and translational motion.

Helmholtz's paper came to the attention of Peter Guthrie Tait, Professor of Natural Philosophy at Edinburgh University, who devised a splendid means of illustrating the vortex theory by the use of smoke rings (see Fig. 9ᵃ). This he demonstrated to Sir William Thomson in January 1867. When two rings were propelled in the same direction with their centers in the same line and their planes perpendicular to this line, the leading ring expanded and moved more slowly. The pursuing ring, on the other hand, contracted and moved faster, eventually moving through the other ring. The process then repeated itself. When, however, two rings were projected toward one another along

13 We are about to consider Thomson's work. For Tait see [P. G. Tait and Balfour Stewart], *The Unseen Universe* (New York, 1875), pp. 95–97.
14 H. von Helmholtz, "Ueber Integrale der hydrodynamischen Gleichungen welche den Wirbelbewegungen entsprechen," *J. für die reine und angewandte Mathematik*, 55 (1858): 25–55. I have not thought it necessary to give an account of the mathematical foundations of the vortex atom theory. Such is to be found in Maxwell's "Atom." A brief sketch of these foundations is to be found in Robert H. Silliman, "William Thomson: Smoke rings and nineteenth-century atomism," *Isis*, 54 (1963): 461–74. I am indebted to Silliman for the account of the background to Thomson's first paper given in this and the next paragraph.

a straight line, both of them expanded and moved more and more slowly, never colliding. Projected toward one another at an oblique angle, the rings glanced off each other without coming into actual contact and began to vibrate violently. Finally, the individual smoke rings resisted all efforts to cut them with a knife—the rings simply moving away from, or wriggling around, the knife.

One month after witnessing this remarkable demonstration, Thomson read a paper on vortex atoms before the Royal Society of Edinburgh. He commences by severely attacking the atom of the chemists:

> The only pretext seeming to justify the monstrous assumption of infinitely strong and infinitely rigid pieces of matter, the existence of which is asserted as a probable hypothesis by some of the greatest modern chemists in their rashly-worded introductory statements, is that urged by Lucretius and adopted by Newton—that it seems necessary to account for the unalterable distinguishing qualities of different kinds of matter.[15]

With the Lucretian atom the properties of matter are explained by attributing these properties to the atom itself. Thus in the kinetic theory of gases the atoms have been assumed elastic in order to explain the elasticity of gases. Every other property of matter has similarly required "an assumption of specific forces pertaining to the atom."

It should be noted that Maxwell could ignore these objections and in 1873 declare himself a follower of Lucretius. Likewise, Maxwell could ignore a further objection of Thomson's to the Lucretian atom. The latter explains that the "dynamical" theory of spectra requires that "the ultimate constitution of simple bodies" should have one or more fundamental periods of vibration, "as has a stringed instrument of one or more strings, or an elastic solid consisting of one or more tuning-forks rigidly connected." (Once again we encounter the analogy with sound.)

> To assume such a property in the Lucretian atom is at once to give it that very flexibility and elasticity for the explanation of which, as exhibited in aggregate bodies, the atomic constitution was originally assumed. If, then, the hypothesis of atoms and vacuum imagined by Lucretius and his followers to be necessary to account for the flexibility and compressibility of tangible solids and fluids were really necessary, it would be necessary that the molecule of sodium, for instance, should be not an atom, but a group of atoms with void space between them. Such a molecule could not be strong and durable, and thus it loses the one recommendation which has given it the degree of acceptance it has had among philosophers.[16]

---

[15] Sir W. Thomson, "On vortex atoms," *Proc. Royal Soc. Edinburgh*, 6 (1867): 94–105, and *Phil. Mag.*, 34 (1867): 15–24; in *Mathematical and Physical Papers by the Right Honourable Sir William Thomson, Baron Kelvin* (6 vols.; Cambridge, 1882–1911), 4: 1–12; p. 1.

[16] *Ibid.*, p. 3.

Maxwell could agree that the Lucretian atom was incapable of producing a spectrum and so was driven to the creation of a molecular theory of spectra. To Thomson such a theory was impossible, for the reasons just given. However, as experiments with smoke rings had indicated, the vortex atom has "perfectly definite modes of vibration, depending solely on that motion the existence of which constitutes it."

There is the problem, however, that even for a simple vortex ring the analytical difficulties involved in finding the fundamental modes of vibration are formidable. In 1867 Thomson had only attempted to find the solution for an infinitely long, straight, cylindrical vortex. He gave one "very simple" result:

Let such a vortex be given with its section differing from exact circular figure by an infinitesimal harmonic deviation of order i. This *form* will travel as waves round the axis of the cylinder in the same direction as the vortex rotation, with an angular velocity equal to $(i-1)/i$ of the angular velocity of this rotation. Hence, as the number of crests in a whole circumference is equal to i, for an harmonic deviation of order i there are $(i-1)$ periods of vibration in the period of revolution of the vortex. For the case $i = 1$ there is no vibration, and the solution expresses merely an infinitesimally displaced vortex with its circular form unchanged. The case $i = 2$ corresponds to elliptic deformation of the circular section; and for it the period of vibration is, therefore, simply the period of revolution. These results are, of course, applicable to the Helmholtz ring when the diameter of the approximately circular section is small in comparison with the diameter of the ring, as it is in the smoke rings.[17]

Thus here vibrations are really rotations—a most important innovation.

Thomson continues that the lowest fundamental modes of the two kinds of transverse vibrations of a ring, such as the vibrations seen with smoke rings, must be much "graver" (more important) than the elliptic vibration of section. Unfortunately, these two kinds of transverse vibrations are not specified. But at least we learn from Tait that the vortex rings vibrate about their circular form as about a position of stable equilibrium.[18]

Thomson next, and for the first and only time, gives a concrete example of how spectral appearances may be explained on the vortex atom theory:

It is probable that the vibrations which constitute the incandescence of sodium-vapour are analogous to those which the smoke rings . . . exhibited; and it is, therefore, probable that the period of each vortex rotation of the atoms of sodium-vapour is much less than 1/525 of the millionth of the millionth of a second, this being approximately the period of vibration of the yellow sodium

17 *Ibid.*, p. 4.
18 P. G. Tait, *Lectures on Some Recent Advances in Physical Science with a Special Lecture on Force* (3d ed.; London, 1885), pp. 297, 299.

light. Further, inasmuch as this light consists of two sets of vibrations coexistent in slightly different periods, equal approximately to the time just stated, and of as nearly as can be perceived equal intensities, the sodium atom must have two fundamental modes of vibration, having those for their respective periods, and being about equally excitable by such forces as the atom experiences in the incandescent vapour. This last condition renders it probable that the two fundamental modes concerned are approximately similar (and not merely different orders of different series chancing to concur very nearly in their periods of vibration). In an approximately circular and uniform disk of elastic solid the fundamental modes of transverse vibrations, with nodal division into quadrants, fulfill both the conditions. In an approximately circular and uniform ring of elastic solid these conditions are fulfilled for the flexural vibrations in the plane, and also in its transverse vibrations perpendicular to its own plane. But the circular vortex ring, if created with one part somewhat thicker than another, would not remain so, but would experience longitudinal vibrations round its own circumference, and could not possibly have two fundamental modes of vibration similar in character and approximately equal in period. The same assertion may, it is probable, be practically extended to any atom consisting of a single vortex ring, however involved, as illustrated by those of the models shown to the Society which consisted of only a single wire knotted in various ways. It seems, therefore, probable that the sodium atom may not consist of a single vortex line; but it may very probably consist of two approximately equal vortex rings passing through one another like two links of a chain. It is, however, quite certain that a vapour consisting of such atoms, with proper volumes and angular velocities in the two rings of each atom, would act precisely as incandescent sodium vapour acts—that is to say, would fulfill the "spectrum test" for sodium.[19]

Thus we see how Thomson is led to postulate the existence of an atom consisting of two linked vortex rings. This is the direction in which the theory will later be developed, though not by Thomson himself.

In the years to follow, Thomson worked on the mathematical development of the vortex theory and among other things proved in 1880 that the vibrations of a columnar vortex are indeed what in 1867 he had said they would be.[20] Other aspects of this work are not without interest to us here. For example, in his 1875 paper entitled "Vortex statics" we find drawings of not only two (Fig. 9b), as in the case of sodium, but also of three vortex rings linked together (Fig. 9c). Furthermore, instead of a circular vortex unit we can have a unit knotted upon itself as in Figures 9d, e, and f.[21] All of these structures are just as permanent as the single circular vortex rings.

This then was the extent of the relevant physical and spectroscopic

19 Sir W. Thomson, "On vortex atoms"; in *Mathematical and Physical Papers*, 4: 4–5.
20 Sir W. Thomson, "Vibrations of a columnar vortex," *Phil. Mag.*, 10 (1880): 155–68; in *Mathematical and Physical Papers*, 4: 152–65; p. 165.
21 Sir W. Thomson, "Vortex statics," *Proc. Roy. Soc. Edinburgh*, 9 (1878): 59–73; in *Mathematical and Physical Papers*, 4: 115–28.

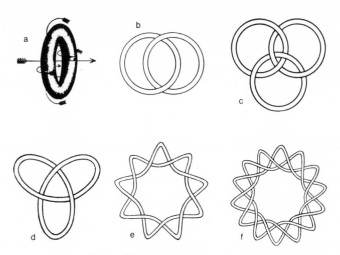

Fig. 9. Illustrations of Vortex Atoms

aspects of the vortex atom theory of matter when in 1875 Maxwell declared it the most satisfactory atomic theory yet advanced. In enumerating the positive attributes of the vortex atom, Maxwell begins with its quantitative permanence as regards volume and strength. It is also qualitatively permanent in regard to complexity, whether it be a knotted or a linked atom. At the same time it is capable of "infinite" changes of form, and may execute vibrations of different period as atoms are known to do. And the number of essentially different "implications"[22] of vortex rings may be great without supposing the degree of complexity to be equally great.

But to Maxwell,

the greatest recommendation of this theory, from a philosophical point of view [and here we have echoes of Thomson, whose objection to the Lucretian atom Maxwell had hitherto chosen to ignore] is that its success in explaining phenomena does not depend on the ingenuity with which its contrivers "save appearances," by introducing first one hypothetical force and then another. When the vortex atom is first set in motion, all its properties are absolutely fixed and determined by the laws of motion of the primitive fluid, which are fully expressed in the fundamental equations. The disciple of Lucretius may cut and carve his solid atoms in the hope of getting them to combine into worlds; the follower of Boscovich may imagine new laws of force to meet the requirements of each new phenomenon, but he who dares to plant his feet in the path opened up by Helmholtz and Thomson has no such resources. His primitive fluid has no other properties than inertia, invariable density, and perfect mobility, and the method by which the motion of this fluid is

---

22 In nineteenth-century usage—"the condition of being involved, entangled, twisted together, intimately connected or combined."

to be traced is pure mathematical analysis. The difficulties of this method are enormous, but the glory of surmounting them would be unique.[23]

It was more the promise, rather than the actual scope and power, of the "infant" theory of vortex atoms which at this point recommended it to Maxwell. But no matter, as far as British spectroscopy was concerned the rout of the Lucretian and Boscovichean atoms was almost complete. Thomson, Maxwell, and Tait all recognized that even in its present undeveloped state the vortex atom theory faced certain serious difficulties,[24] but in spite of these, and here we will find a contrast with continental European scientists, they embraced the theory and worked with it. And their influence was such that in Britain and in America the vortex theory of matter was to reign for the quarter of a century down to about 1900.[25]

But then, we ask, were all of the British not in a contradictory position in accepting the vortex atom on the basis of the Kundt–Warburg result? The answer depends on the attitude to the equipartition principle. If the principle were completely accepted the answer would be yes. And if it were rejected outright then there would be no reason for even considering the Kundt–Warburg result to begin with, and again the answer would be yes. If, however, a modified equipartition principle such as that first given in 1876 and again in 1893 by the Reverend Henry William Watson were accepted, then the answer would be no.

Watson pointed out that underlying the specific-heats difficulties of the kinetic theory was the assumption that any disturbance, such as heating, is instantaneously communicated equally to all degrees of freedom of the system.[26] And he suggested that this, however, might not be so. It might be said that the first effect of a disturbance is to alter the translational energy of the molecule, the latter being so constituted as to require a finite or even long interval of time for complete equipartition. Thus the number of effective degrees of freedom during the period in which specific heats are being determined would be smaller than if complete equipartition were reached instantaneously. In this way the Kundt–Warburg result could be reconciled with a vibrating atom. But

---

23 J. C. Maxwell, "Atom," pp. 471–72.
24 In "Atom," for example, Maxwell discusses the difficulties of explaining mass and gravitation on the vortex atom theory. *Ibid.*, p. 472 ff.
25 For the adoption of the vortex atom theory in America see Silliman, "William Thomson: Smoke rings and nineteenth-century atomism."
26 H. W. Watson, *A Treatise on the Kinetic Theory of Gases* (2d ed., Oxford, 1893), pp. 80–87.

then this latter did not necessarily have to be a vortex atom, it could be the aggregate of elastically connected ultimate atoms known as a chemical atom.

The dean of British spectroscopists, Arthur Schuster, was well aware of this last point, and perhaps this is one reason—the other being that he saw the contradiction in Maxwell's position, as Maxwell himself later did—why he seems never to have accepted the vortex atom theory of matter. In an address on "modern spectroscopy" given at the Royal Institution in 1881 he never once mentions vortex atoms. He does mention the Kundt–Warburg experiment, but only to attack its interpretation according to kinetic theory.

Schuster explains that "modern spectroscopy" is "the union of the old [spectroscopy] with the modern ideas of the dynamical theory of gases, and includes the application of the spectroscope to the experimental investigation of molecular phenomena, which without it might forever remain matters of speculation or of calculation."[27] To Schuster a body is made up of atoms which "are hardly ever, perhaps never, found in isolation."[28] In a molecule two or more atoms are bound together and do not part company so long as the physical state of the molecule remains the same. Within each molecule the atoms move relative to one another while each atom "has its own internal movement."[29]

The atoms have parts and vibrate, but for Schuster every spectrum is nevertheless produced by a molecule. (We are reminded of the considerations of section 4 of chapter II.) This is true even in the case of mercury vapor, as is seen when Schuster speaks of "certain" conclusions of the molecular theory of gases which "are at present absolutely contrary to fact."

A spectroscopist . . . who is acquainted with the mercury spectrum and all the changes in that spectrum which can take place, feels more than skeptical when he is told that the molecule of mercury contains only one atom, which neither rotates nor vibrates.

Nor can it be of advantage to science to pass silently over the difficulty, or to neglect it as unessential, as is often done by modern writers. The late Professor Maxwell, at least, was well aware of its importance, and has often expressed in private conversation how serious a check he considered the molecular theory of gases to have received.[30]

---

27 A. Schuster, "The teachings of modern spectroscopy," *Popular Scientific Monthly*, 19 (1881): 466–82; p. 469.
28 *Ibid.*
29 *Ibid.*
30 *Ibid.*, p. 481.

With Schuster's view the grand theoretical goal is still knowledge of the chemical forces binding together the atoms of a molecule and not knowledge of the structure of atoms.

I believe no other line of investigation to be as hopeful (in gaining a deeper insight into the nature of chemical forces) as the one which examines directly the vibrations of the molecules which take place under the influence of these chemical forces. If we could find a connection between the vibrations of a compound molecule and the vibrations of the simpler elements which it contains, we should have made a very decided step in the desired direction.[31]

However, within Britain there was a line of people commencing with Sir William Thomson who associated spectra primarily with the atom. Such a person was William Hicks, one of the few people who made positive contributions to the mathematical development of the vortex theory. Writing in 1883 Hicks says that, "The vortex atom theory has its most interesting connection with the explanation of the spectral lines of the elements. These lines, *so far as they depend on the vibrations of single atoms,* might arise from several different kinds of vibrations of the form of the ring."[32]

Hicks developed the theory of a hollow vortex ring which he explains may have some of the following features:

1. Deformations of the circular axis. As was shown by Thomson these must be such that the axis at any time is deformed into a helix wound on the surface of a tore, or the ring is twisted.[33]
2. Waves running round the surface of the ring, so that any cross-section is crimped into small elevations and depressions, and the ring itself is fluted.
3. Vibrations of the aperture. This is a particular case of (2).
4. Pulsations of the hollow interior of the vortex ring, whereby the volume of the hollow is periodically altered.
5. Swellings of the ring, traveling in one direction or the other round the ring, so that the ring seems to be braded.[34]

Hicks was of the opinion that if relations were to be found between spectral lines—and we recall that 1883 was two years before Balmer announced his hydrogen series formula—then it would be in classes 1, 2, and 5. However, he points out that from Thomson's investigations it would appear that with a solid tore the periods of class 1 would be

31 *Ibid.*, p. 475.
32 W. Hicks, "On the steady motion of a hollow vortex," *Proc. Royal Soc.*, 35 (1883): 304–8; p. 306.
33 A tore is a solid generated by the revolution of a circle or other conic about any axis—for example, a solid ring of circular or elliptic section.
34 W. Hicks, "On the steady motion of a hollow vortex," p. 306.

temperature dependent. But this is not true of class 2, as Hicks himself has found.

While at first this must have appeared as a promising approach in the quest for numerical relationships between spectral lines, it was in fact unproductive. As we have seen, the vortex atom theory of matter played no part in the discovery of spectral series formulae. These were discovered by continental spectroscopists—Balmer, Rydberg, Kayser, and Runge—who never adopted the vortex atom theory. We might say that Stoney had a part in Balmer's success, but Stoney's theory of spectral relationships was formulated before the vortex atom theory was in vogue among British spectroscopists.

As will be seen in a later section, Stoney constructed a second theory of spectral relationships on the basis of the vortex atom theory. It is not certain that he chose any of the five classes of periodic motions listed by Hicks, but then the vortex atoms were richer in periodic motions than even Hicks allowed. To illustrate this potential, to prepare the way for the later consideration of Stoney's work, and to show how the vortex atoms of the various elements were imagined physically, we turn now to the work of another outstanding figure in the development of the vortex atom theory of matter, Joseph John Thomson. It will also be helpful later to know that Thomson, as most of the British, was imbued with the theory of vortex atoms.

In 1883 Thomson published *A Treatise on the Motion of Vortex Rings* for which he had been awarded the Adams Prize at the University of Cambridge in the previous year. Included in the published work, though not in the original essay, is "a sketch of a vortex atom theory of chemical action." In this we see Thomson attempting to construct the various atoms of the periodic table.

Unfortunately, Thomson never uses diagrams and his purely verbal descriptions are not always easy to follow. At all times he considers vortex rings whose cross-sectional diameter is small as compared with the aperture of the ring. He shows that if two rings are linked then they spin round one another, and gives the following description of how we are to conceive the linkage:

To get a clear conception of the way the vortices, supposed for the moment of equal strength, are linked, we may regard them as linked round an anchor ring whose transverse section is small compared with its aperture, the manner of linking being such that there are always portions of the two vortices at opposite extremities of a diameter of a transverse section of the anchor ring. The shortest distance between pieces of the two vortices is then approximately constant, and equal to the diameter of the transverse section of the anchor ring.[35]

---

[35] J. J. Thomson, *A Treatise on the Motion of Vortex Rings* (London, 1883), p. 78.

In addition to the motion of one ring about the other, any particle on the surface of a ring rotates about the core axis of that ring. Thus the compounded motion of the particle is an epicyclic one.

Thomson demonstrates that any number, not exceeding six, of equal vortices may be linked in this manner to form a unit which is in steady stable motion. If more than six are linked together the unit is unstable. In a stable configuration the vortices are arranged so that their core axes pass through the angular points of a regular polygon inscribed in the circular transverse section of the anchor ring.

To have a stable configuration it is not necessary that each vortex ring should be single. One can have composite vortex rings, provided the distances between their components are small compared with the sides of the polygon, at the angular points of which the vortices are situated, and provided that the sum of the strengths of the components is the same as the strength of the single vortex ring, which they are supposed to replace. Thomson refers to the systems of vortices placed at the angular points of the polygon as the primaries, and the component vortex rings of these primaries as the secondaries of the system.

Thomson then supposes that the atoms of the different chemical elements are made up of vortex rings all of the same strength.[36] He says that some of the elemental atoms consist of only one of these rings, others of two of the rings linked together, and yet others of four, five, or six linked vortices. But he does not say how many ultimate rings there may be.

The behavior of two vortex rings projected one after the other along the same straight line—which behavior, we recall, Tait had demonstrated with his smoke rings—is taken to illustrate the process of chemical combination.[37] The rings repeatedly move through one another. "We may suppose that the union or pairing in this way of two vortex rings of different kinds is what takes place when two elements of which these vortex rings are atoms combine chemically." Nevertheless, it is difficult to picture how two complex atoms would combine chemically —what would pass through what?

Regarding spectra, Thomson merely remarks in the closing paragraph that:

According to the view we have taken, atomicity [valency] corresponds to complexity of atomic arrangement, and the elements of high atomicity consist of more vortex rings than those whose atomicity is low; thus high atomicity corresponds to complicated atomic arrangement, and we should expect to find the spectra of bodies of low atomicity much simpler than those of high. This seems

36 *Ibid.*, p. 119.
37 *Ibid.*, p. 114.

to be the case, for we find that the spectra of Sodium, Potassium, Lithium, Hydrogen, Chlorine which are all monad elements, consist of comparatively few lines.[38]

Thus while some progress was made in developing the vortex atom theory of matter during the eight years down to 1883, the same cannot be said for the corresponding theory of spectra. Nor was any progress made in the succeeding seven years down to 1890 when, as we shall see below, the mechanical theory began to be replaced by the electromagnetic theory of spectra. Nevertheless, during the last quarter of the century most of the British regarded spectral emitters as being single or combined vortex atoms.

It is now time to consider how these emitters were conceived in certain continental European countries.

2. The Atomic-Molecular Theory of Spectra in Continental Europe

The first continental response to the Kundt–Warburg result came from France, a country which nevertheless was to contribute relatively little to either kinetic theory or physical spectroscopy in the last quarter of the nineteenth century.

In 1876 Antoine Yvon Villarceau had a paper on kinetic theory which incorporated the Kundt–Warburg result read before the Académie des Sciences.[39] The paper drew a response from the eminent chemist, Marcelin Berthelot, and it is this which is of interest here because Berthelot was, so to speak, looking at physical views from the outside.[40] He mentions the statements, made by physicists, that the ratio of the specific heats of a gas composed of material points would be 1.66, but that if allowance is made for internal motions the ratio would be less than this. And he correctly describes them as "hypotheses, envisaged as pure representative abstractions." Berthelot is most reluctant to accept the conclusion that the molecules of mercury vapor are material points. To him many other demonstrations in addition to that of Kundt and Warburg would be required before this could be established.

The only thing which the hypothesis of such a monatomic matter has in common with the conceptions of the chemists, he says, is the name. We recall that, according to Berthelot, chemists considered only the

38 *Ibid.*, p. 124.

39 A. Yvon Villarceau, "Note sur les déterminations théorique et expérimentale du rapport des deux chaleurs spécifiques, dans les gaz parfaits dont les molecules seraient monoatomiques," *Comptes Rendus*, 82 (1876): 1127–28.

40 M. Berthelot ["Sur l'existence réelle d'une matière monatomique"], *Comptes Rendus*, 82 (1875): 1129–30.

relationship between the weights of "molecules" which combine with or substitute one another, their atoms being defined by the minimum values met with in these relationships. Thus Berthelot adopted in chemistry an approach equivalent to that adopted by Maxwell in physics; each was well aware of what was hypothesis and what fact. But there were many chemists and physicists who readily accepted mercury vapor, for example, as monatomic—as we have seen.

To Berthelot the concepts of an indivisible yet extended and continuous atom, and of an atom endowed with mass yet reduced to a material point, seemed to be self-contradictory—"as many philosophers have not ceased to think since the time of the Greeks and of Boscovich, first promoters of these hypotheses."[41] But what then is the nature of an atom? Berthelot is unable to say.

Another Frenchman, Charles Simon, an astronomer and mathematics teacher, regarded the Kundt–Warburg result in a way similar to Bosanquet; namely, that it suggested that the spectral energy of an atom is very small compared to the thermal energy. And like Bosanquet, Simon also used this inference to obtain agreement between the calculated and observed values of the specific heats of such gases as hydrogen, oxygen, and nitrogen. Simon regards these not as diatomic but as tetratomic. He thinks of the four atoms in a molecule as occupying the apices of a regular tetrahedron, whose edge is much greater than the diameter of an atom, and the interior of this tetrahedron as filled with free or condensed ether. In taking account of the rotation of each elementary tetrahedron about its center of gravity and regarding the vibrations of the atoms as "null or insensible," he finds the value of the ratio of the specific heats to be 1.4—giving good agreement with experiment. "From this accord between calculation and experiment, it seems to result that the internal vibrations, neglected in the calculation, are really negligible."[42] Thus we can see today that while unsuccessful with his molecular model Simon, as Bosanquet, was correct in his general conclusion.

Simon's value for the ratio of the specific heats of his tetratomic molecules was disputed soon afterward by Boltzmann, whose interesting reaction to the Kundt–Warburg result we now consider. Boltzmann, we recall, had conceived gas molecules as being "aggregates of material points (atoms), held together by forces which are arbitrary functions of the separation of the atoms."[43] However, Boltzmann did not think

41 *Ibid.,* p. 1130.
42 C. Simon, "Sur le rapport des deux chaleurs spécifiques," *Comptes Rendus,* 83 (1876): 726–28; p. 728.
43 L. Boltzmann, "Über die Natur der Gasmoleküle," *Berichte Akad. Wissen. Wien,* 74 (1876): 553–60; in *Wissenschaftliche Abhandlungen,* 2: 103–10.

that it necessarily followed from the Kundt–Warburg result that "a molecule of mercury is truly a mathematical point," a fact which he considered to be refuted by the mercury spectrum. Rather the result showed that the "molecules" of mercury vapor behave almost as material points, or, if one prefers, as elastic spheres. To Boltzmann the latter, as the former, have only three degrees of freedom.

Boltzmann's reaction was to "freeze" his molecules and regard the mercury molecule as an elastic sphere composed of both material and etherial atoms. The mercury spectrum is then explained as being produced by the vibrations which occur during and for a very short time after each molecular collision, in a similar way to the acoustical situation where a click is emitted when two billiard balls collide. Thus both spectral and kinetic theory requirements are met. However, the generally received theory of spectra has been sacrificed—the vibrations of the molecule are not communicated to the ether during the free path following a collision, but during the collision and for a very short time thereafter.

Boltzmann then considers the case of a molecule composed of two material points, or two elastic spheres, rigidly connected. As Bosanquet also pointed out, this type of molecule possesses five degrees of freedom—three of translation and two of rotation. The ratio of the specific heats is calculated to be 1.4, a value which is close to that determined experimentally for "air and most remaining simple gases." The same value is obtained for a molecule consisting of any number of rigidly connected material points or elastic spheres lying in a straight line. On the other hand if the components do not lie in a straight line, or more generally if the molecule is a rigid body which is not a solid of revolution, then the number of degrees of freedom is six and the ratio of the specific heats 1.33—a value which had been found experimentally for some gases. Boltzmann points out that this is the value which Simon ought to have calculated for his rigid tetratomic molecule.

Boltzmann, however, has still to account for the spectra of polyatomic molecules and he says that it "is of course obvious, that the gas molecules cannot be absolutely rigid bodies." What was said about the mercury molecules is also valid here—spectral radiation is emitted during molecular collisions. The molecules are supposed to penetrate into one another to a small distance directly dependent upon the velocities of impact, just like elastic spheres or elastic ellipsoids of revolution which in the instant of collision flatten one another a little, but then recover their original shape.

It is of interest to compare for a moment the reactions of Maxwell and Boltzmann to the Kundt–Warburg result. Maxwell, we recall, was initially in the paradoxical position of both accepting and rejecting the result, although according to Schuster he afterward tended more toward the latter and claimed kinetic theory to be somehow incomplete. On the other hand, Boltzmann seizes upon the result as providing a way out of the serious specific-heats difficulties of kinetic theory, and at the same time does not hesitate to make the changes he requires in spectral theory. Boltzmann's reaction and Maxwell's initial reaction reveal, as would be expected, a strong commitment to kinetic theory. But spectroscopic requirements also weighed heavily with Maxwell and eventually exerted their influence on his thought. Before the Kundt–Warburg result Maxwell had shown considerable interest in spectral theory, whereas Boltzmann had not.

We are fortunate in knowing Maxwell's assessment of Boltzmann's position. Maxwell criticized it in 1877 at the end of his review in *Nature* of Watson's treatise on the kinetic theory of gases.[44] Outlining Boltzmann's theory Maxwell first attacks it in relation to the question of specific heats. Let us try, he says, to construct a rigid-elastic body such as Boltzmann envisages.

It will not do to take a body formed of continuous matter endowed with elastic properties, and to increase the coefficients of elasticity without limit till the body becomes practically rigid. For such a body, though apparently rigid, is in reality capable of internal vibrations, and these of an infinite variety of types, so that the body has an infinite number of degrees of freedom.

However, that is not all. The same objection applies to all atoms, such as Thomson's vortex atoms, constructed of continuous nonrigid matter. Such atoms would soon convert all of their translational energy into internal energy, and the specific heat of a substance composed of them would be infinite. Maxwell, it seems, would conceal nothing in his quest for scientific truth. He fully appreciated "that state of thoroughly conscious ignorance which is the prelude to every real advance in knowledge."[45]

Maxwell's second criticism is made in relation to spectral requirements. He explains that a truly rigid-elastic body is one whose encounters with similar bodies take place as if both were elastic, but which is not capable of being set into internal vibration. We must take a perfectly rigid body and endow it with the power of repelling all other bodies, but only when they approach within a very short distance of it, and then so strongly that under no circumstances whatever can any body come into actual contact with it.

To Maxwell this was the only constitution possible for a rigid-elastic

44 J. C. Maxwell, "The kinetic theory of gases," *Nature*, 16 (1877): 242–46.
45 *Ibid.*, p. 245.

body. And having obtained it he thought the best thing to do was to get rid of the rigid nucleus and substitute a Boscovichean atom—"A mathematical point endowed with mass and with powers of acting at a distance on other atoms."

However, Maxwell was well aware that Boltzmann's molecules were not absolutely rigid. "He [Boltzmann] admits that they vibrate after collisions, and that their vibrations are of several different types, as the spectroscope tells us. But still he tries to make us believe that these vibrations are of small importance as regards the principal part of the motion of the molecules." It is interesting to see the two leaders of kinetic theory differing here. Part of the explanation of this would seem to be that Maxwell was more committed to spectral theory than Boltzmann.

Maxwell explains that if one were to suppose that at ordinary temperatures the collisions are not severe enough to produce any internal vibrations, and that vibrations occur only at temperatures like that of the electric spark, when measurements of specific heats cannot be made, one might, then, perhaps, reconcile the spectroscopic and specific heats results. However, the fixed positions of the bright lines of a gas show that the vibrations are isochronous, and therefore that the forces which they involve vary directly as the relative displacements. And since this is the nature of the forces, all impacts, however slight, will produce vibrations. Maxwell adds that even at ordinary temperatures certain gases, as iodine vapor, display absorption bands, which indicates that the molecules are set into internal vibrations by the incident light. Thus, there is no escaping the fact that the internal vibrations must take up their due proportion of energy according to their number of degrees of freedom.

Maxwell, nevertheless, could not see his way out of this dilemma, in which he was also caught. Specific heat requirements demanded only a small number of degrees of freedom, and spectral requirements a greater but nevertheless finite number; yet both Boltzmann's atomic model and the vortex atom had an infinite number of degrees of freedom. Thus only two years after strongly endorsing the vortex atom Maxwell indicated that it was far from satisfactory. But as I have stated earlier the vortex atom, in spite of its numerous difficulties, was generally accepted in Britain during the last quarter of the nineteenth century.

Yet another atomic-molecular explanation of spectra was given in 1878, this time by Eilhard Wiedemann, Extraordinary Professor of Physics at Leipzig.[46] Wiedemann explains that J. Stefan and J. D. van

---

[46] E. Wiedemann, "Untersuchungen über die Natur der Spectra (1. Theorie 2. Spectra gemischter Gase)," *Ann. der Physik*, 5 (1878): 500–24; in *Phil. Mag.*, 7 (1879): 77–95.

der Walls hold that attractive forces exist between gaseous molecules and extend to distances which are great in comparison with the diameter of a molecule. At minute distances, however, the molecules must repel one another, since otherwise they would not recoil after a collision. It is Wiedemann's view that the repulsive forces have their origin in "the envelopes of ether surrounding the molecules" and that they "must diminish more rapidly with increasing distance than the attractive forces." We recall that this molecule is similar to that mentioned by another German, Clausius, much earlier. Indeed we find this molecule or atom employed by many other German physicists after Clausius, and it would thus seem to have been in Germany what the vortex atom later was in Britain.[47]

Wiedemann holds that at the highest possible temperatures the molecules of a rarefied gas are dissociated into atoms. "When these collide only oscillatory motions will occur, since, according to the Kundt-Warburg experiment on the specific heat of mercury vapour, as well as the theoretical reflections of Maxwell, Watson, and Boltzmann, the *vis viva* of rotational motion is *nil* in monatomic molecules." But did theory not also say that the energy of vibration should likewise be zero? Wiedemann ignores this and says that these atomic vibrations are the source of line spectra. Thus with Wiedemann the Kundt–Warburg result did not entail a change of atomic model, and this would seem to be true of German physicists in general. Part of the reason must be that they could not accept vortex atoms and there was simply no other model to which they could turn.

With the German model, at least initially, it was not the material atom itself but its ether envelope which was supposed to vibrate. Wiedemann says that at low temperatures the amplitudes of the motions of the individual ether particles are supposed small. A limited number of different rays, whose wavelengths depend upon the arrangement of the ether particles and the forces acting between them and the

---

[47] See, for example: E. Lommel, "Ueber Fluoreszenz," *Ann. der Physik,* 143 (1871): 26–51; and A. Handl, "Notiz ueber die absolute Intensität und Absorption des Lichts," *Berichte Akad. Wissen. Wien.,* 64 (1872), 2: 129–32. "Molecules" having ether envelopes were postulated by O. F. Mosotti in 1837 with a view to explaining, among other things, radiation—"On the forces which regulate the internal constitution of bodies," *Scientific Memoirs,* ed. Richard Taylor, 1 (1837): 448–469. This type of atom or molecule was accepted in some quarters in Britain during the 1830s and 1840s—see David M. Knight, *Atoms and Elements* (London, 1967), pp. 60–70. The vortex atom of W. J. Macquorn Rankine was actually such an atom, the vortices occurring in the ether envelope. Thus Rankine's vortex atom was different from William Thomson's vortex atom, a fact which has been overlooked by the editors of *Isis* who illustrate Edward E. Daub's discussion of Rankine's atom ("Atomism and thermodynamics," *Isis,* 58 (1967): 293–303) with sketches of Thomson's atoms.

material atom, are emitted. On raising the temperature certain harmonics, which again depend upon the disposition of the ether envelopes, will be added to the fundamental wavelengths.

Wiedemann then gives the required qualitative explanations of the better known spectral phenomena—as, for example, the broadening of spectral lines when the pressure of the emitting gas is increased. In this case the individual ether envelopes can vibrate undisturbed for only a brief time, since they are for the most part within the spheres of action of other atoms. The radius of a sphere of action is the length to which, measured from the center of the atom or molecule, the ether is in a state different from that of the ether of free space.

In the vortex atom theory the difference of state was a difference of motion, but here we are not told the nature of the difference. It is also worth noting that Maxwell would have felt compelled to attribute three degrees of freedom to each ether particle in Wiedemann's atomic envelopes.

Since the line spectra of chemically similar elements, such as sodium and potassium, are similar, Wiedemann concludes that the configurations of the associated ether envelopes, or the forces acting upon them, must also be similar. Among similar spectra a displacement of corresponding lines toward the violet signifies an increase of the attractive force between the atom and the ether envelope.

Thus Wiedemann was able to make the German model appear at least qualitatively satisfactory. And so in Germany as well as in Britain —and we might also add Austria for Boltzmann's molecules are composed of etherial, as well as material, ultimate atoms—the luminiferous ether was essential not only for the transmission but also for the production of spectral radiation. No continental nation, however, would follow Britain in regarding ponderable matter as a state of motion of the ether.

Writing in *Nature* in 1880 about the vortex atom theory, Samuel Tolver Preston, a British telegraph-engineer, says:

Much apprehension would seem to exist in regard to the physical side of the theory, especially in Germany, where the mathematical investigations out of which it sprung had their origin. Some appear to be unable to conceive how motion should take place in a material substance continuously filling space, losing sight of the fact that the liquid outside the atoms plays the part of a void (in so far as it cannot appeal to our senses)—or it is only the atoms which affect our perceptions. Others fail totally to appreciate the physical side of the theory, and seem to think that it involves arbitrary postulates, whereas the main peculiarity of the theory is its freedom from positive assumptions, inasmuch as the theory evolves all the properties of matter out of the *motion* of a material substance . . .. The fact seems to be overlooked that if we renounce the occult quality of *rigidity* in the atom, we have no other resource than a *liquid* (i.e. a substance without rigidity). Much of the misunderstanding on the subject may no doubt be due to the scarcity of the literature relating to it, and the extreme brevity and absence of detail or attempts to assist the conceptions regarding the physical side of the theory.[48]

Indeed, in 1880 the only writings in English dealing with the vortex theory which Preston could cite were those of Thomson, Tait, and Maxwell.

In discussing the vortex atom theory in his *Lectures on Some Recent Advances in Physical Science* (1876), Tait had said that

to investigate what takes place when one circular vortex atom impinges upon another, and the whole motion is not symmetrical about an axis, is a task which may employ perhaps the lifetimes, for the next two or three generations, of the best mathematicians in Europe; unless, in the meantime, some mathematical method, enormously more powerful than anything we at present have, be devised for the purpose of solving this special problem. This is no doubt a very formidable difficulty, but it is the only one which seems for the moment to attach to the development of this extremely beautiful speculation; and it is the business of mathematicians to get over difficulties of that kind.[49]

As we know, the British proceeded to develop the theory. However, according to Preston, this passage was regarded by certain Germans "as if the difficulty there mentioned were of such a nature as to prevent the practical adoption of the theory."[50]

In Holland the vortex atom theory was criticized and rejected in 1889 by Victor August Julius, Professor of Experimental Physics at Utrecht.[51] The Thomsons had developed the theory for rings whose transverse section is small relative to the diameter of the ring, and Julius objects that no one has as yet succeeded in developing the theory for rings whose transverse sections are not relatively small. Again, no agreement has yet been obtained between wavelengths calculated on the vortex atom theory and observed spectral wavelengths.

Julius adds what he feels to be still stronger arguments against the theory, although they do not involve spectral considerations. First, the vortex atom theory did not afford an explanation of gravitation, as Tait and Maxwell had acknowledged.[52] However, an explanation of gravi-

---

[48] S. T. Preston, "On the physical aspects of the vortex-atom theory," *Nature*, 22 (1880): 56–59.

[49] P. G. Tait, *Lectures on Some Recent Advances in Physical Science,"* pp. 302–3.

[50] S. T. Preston, "On the physical aspects of the vortex-atom theory," p. 57.

[51] V. A. Julius, "Sur les spectres des lignes des éléments," *Annales de L'École Polytechnique de Delft,* 5 (1889): 1–117.

[52] *Ibid.,* pp. 42–43.

tation had been given by Hicks and this Julius had apparently over-looked.[53] But second, and Julius regarded this as the most important objection, in considering a gas composed of circular vortex rings J. J. Thomson had been unable to arrive at a satisfactory explanation of Boyle's law. Finally, there was the difficulty also encountered by J. J. Thomson that when the temperature of a gas consisting of vortex atoms was raised the translational velocity of the atoms decreased. To Julius it seemed inadmissible that this represented reality.

In 1886 Josef Loschmidt had adopted an atom similar to Wiedemann's, and treating the ether envelope as an elastic sphere had attempted to calculate its vibrations.[54] But Julius also rejected this approach, saying he found it difficult to admit that if the atoms have ether envelopes that these envelopes would behave as hollow elastic spheres. He also rejected the idea of the atoms themselves behaving like elastic spheres. In fact Julius was of the opinion that for the moment there was no great prospect of any hypothesis on the nature of the atom leading to an explanation of spectra.

We end this section, as we began it, by considering a French view of gaseous molecules. This was published in 1891. We noted at the beginning of the section that France made no important contributions to the physical phase of spectroscopy in the last quarter of the nineteenth century, so it comes as no surprise that the paper we are about to consider, in spite of the fact that it appears in the *Comptes Rendus,* is of extremely poor quality and quite lacking in originality. The author was Marcel Brillouin, a lecturer in physics at the École Normale Supérieure.[55]

Brillouin considers what he calls three principal hypotheses on the constitution of gaseous molecules in regard to the production of spectra. The first of these, however, had been rejected after the late seventies by most of the leading spectroscopists with the notable exception of Rydberg, who seems never to have been troubled by the difficulty which others, including Brillouin, saw associated with it. The second likewise had become more or less rejected after the early eighties. And the third had never been admitted by competent spectroscopists.

Brillouin's first molecule is a complex one made up of numerous parts. These parts are "true atoms." "The chemical atom of a simple body ought to be considered as an agglomeration of an extremely great,

53 W. Hicks, "On the problem of two pulsating spheres in a fluid," *Proc. Cambridge Phil. Soc.,* 3 (1880): 276–85.
54 J. Loschmidt, "Schwingungen einer elastischen Hohlkugel," *Sitzungsberichte Akad. Wissen. Wien,* 93 (1886): 434–36.
55 M. Brillouin, "Sur le degré de complexité des molecules gazeuses," *Comptes Rendus,* 112 (1891): 575–77.

but limited, number of distinct atoms of another matter." Julius mentions that this idea of the chemical atom or physical molecule was popular at the time with some physicists, and no doubt this included continental European as well as British physicists. Indeed we have seen elsewhere that Boltzmann and Kayser adopted such a view. This, however, is not to put Kayser in the same camp as Rydberg; their different viewpoints were mentioned at the end of the previous chapter. Brillouin says that for this type of molecule it is necessary to allow as many geometric variables as there are lines in the spectrum—several hundred for iron. But as this would mean a "rather great complexity" for a simple gas, as well as for other reasons, Brillouin is inclined to reject this model.

We recall that Stoney had removed the necessity of requiring highly complex molecules by devising his theory which related a harmonic series of spectral lines to a single motion within the molecule, thus requiring a much simpler molecule. It is this latter type of molecule and related theory which Brillouin mentions as the second principal hypothesis, which he considers to be more satisfactory.

The third and final hypothesis had never been admitted by anyone who, like Stoney in 1868, had taken a little time to make numerical calculations. The hypothesis is that the rapid motions of the atoms and molecules through the ether give rise to spectra, much as a rod drawn swiftly through the air produces a sound.

While he rejected the first hypothesis, Brillouin saw no cause to choose between the second and third. Observation, he thought, would eventually dictate the choice.

Thus we have seen that in continental Europe, as in Britain, there were several attempts in the 1870s and 1880s to give a mechanical explanation of spectra at an atomic/molecular level. But there had been considerable criticism among spectroscopists of one another's proposed explanations. And the outcome was that even at a qualitative level there seemed little prospect of a unified view on the origins of spectra emerging. This, then, was the state of affairs when about 1890 the electromagnetic view of spectra began to be advanced as an alternative to the mechanical view.

3. The Electromagnetic Theory of Spectra

About 1888, as is well known, Heinrich Hertz first gave a demonstration of Maxwell's prediction that light is an electromagnetic radiation.[56] It was primarily from this experimental result that spectroscopists

---

[56] H. Hertz, "Ueber Inductionserscheinungen hervorgerufen durch die electrischen Vorgänge in Isolatoren," *Ann. der Physik*, 34 (1888): 273–85; "Ueber die Ausbreitungsgeschwindigkeit der electrodynamischen Wirkungen," *ibid.*, pp. 551–69.

started anew in their search for a theory of spectra. They were also to make use of the idea, originating in electrolysis and given expression by Stoney and Helmholtz, that each atom of matter has an associated "atomic" quantity of electricity—to which Stoney was to give the name "electron."[57] Apparently, the earliest attempt to calculate spectral wavelengths on the basis of the electromagnetic theory of light was made in 1887. But consideration of this belongs more properly to the following chapter.[58] Two years later the possibility of spectral lines having an electrical origin was also considered by Julius.[59] Julius had read a suggestion by Friedrich Zöllner that the theory of vibrational motion of a pair of oppositely charged "electrical atoms" developed by Wilhelm Weber, would probably lead to a theory of the distribution of lines in a spectrum.[60] For the purpose of applying Weber's theory to spectra Julius assumes every chemical atom to consist of two smaller atoms of equal but opposite charge.[61]

Julius proceeds by further assuming, as Stoney had done earlier, that the function expressing the vibrational motion of the pair may be expanded into a Fourier series. Harmonic relationships should then exist between the wavelengths of the lines of any spectrum. But, says Julius, since "the researches of Schuster have shown that these harmonics do not exist, one cannot, to my mind, expect that the vibrational motion of atoms is analogous to that of a couple of electric atoms in the theory of Weber."[62] Julius does not exclude the possibility of every chemical atom being a complex of electric atomic couples of which only the fundamental vibrations, and perhaps some harmonics, would be visible. He does not, however, attempt a mathematical treatment.

A further two years elapsed and another, quite different, electromagnetic theory of spectra was suggested by Hermann Ebert, a *Privatdozent* at Erlangen. In this theory atoms are considered as Hertzian oscillators.[63] For the background to the work we must briefly return

57 G. J. Stoney, "On the physical units of nature," *Phil. Mag.*, 11 (1881): 381–90; "Of the 'electron' or atom of electricity," *ibid.*, 38 (1894): 418–20. H. von Helmholtz, "On the modern development of Faraday's conception of electricity," *J. Chem. Soc.*, 39 (1881): 277–304.
58 See on to p. 201.
59 V. A. Julius, "Sur les spectres des lignes des éléments," pp. 38–40.
60 F. Zöllner, *Principien einer electrodynamischen Theorie der Materie* (Leipzig, 1876), p. xxi.
61 F. Richarz later considered two oppositely charged atoms combined in a molecule and rotating about one another as a spectral emitter—"Ueber die electrischen und magnetischen Kräfte der Atome," *Ann. der Physik*, 52 (1894): 385–416; pp. 407–10.
62 V. A. Julius, "Sur les spectres des lignes des éléments," p. 39.
63 Fürst B. Galitzin was later also to regard the atom as some type of Hertzian oscillator—"Zur Theorie der Verbreiterung der Spectrallinien," *Ann. der Physik*, 56 (1895): 78–99.

to purely mechanical considerations and consider a modification which Wiedemann, who was Ebert's superior at Erlangen, had introduced into the German theory of spectra considered in the previous section. We recall that Wiedemann had held the ether envelopes of atoms and molecules to be the seat of spectral radiation. In 1889, however, he reconsidered this view during the course of a long paper on the mechanics of light.[64] From experiment Wiedemann knows the amount of spectral energy radiated by a given molecule, and he believes that this energy is originally in the molecule as kinetic energy—either of the ether envelope or of the material part of the molecule. He is going to decide between these two alternatives by means of calculation.

Wiedemann can express the energy of a spectral line of given wavelength as function of the weight and of the amplitude of vibration of that part of the molecule, whether etherial or material, producing the radiation. The energy is proportional to the product Ma^2 where M is the mass, and a the amplitude of the motion of the vibrating entity. For a given molecule Wiedemann knows the value of M from the kinetic theory of gases. He also knows the volume of the molecule and apparently the density of the ether, so he can compute the weight of free ether contained in a volume equal to that of the molecule. And for the purpose of the calculation he adopts the hypothesis that this is the weight of ether associated with the molecule.

Wiedemann can now calculate the magnitude of the amplitude, assuming, first, that it is the ether envelope which vibrates, and second, that rather it is the material part which vibrates. These values can then be compared with the diameter of the molecule. In all cases it is found that the amplitude for the ether envelope is about 10^5 times as great as the molecular diameter, which is "wholly unthinkable." Thus the radiation must be emitted by the material part of the molecule. In this case the amplitudes are only a small fraction of the molecular diameter.

It is with this work of Wiedemann in mind that Ebert attempts to construct his electromagnetic theory of spectral radiation. As the source of radiation Ebert assumes electrically charged atoms and molecules which correspond to Wiedemann's mechanically deformable atoms and molecules. The charges on the atoms and molecules may become displaced under the action of electromotive forces, and so the atoms and molecules may become polarized. From a general application of Maxwell's equations Ebert first determines that such atoms and molecules would emit as much spectral energy as had been determined by Wiedemann from purely mechanical considerations. Ebert then considers the electrical behavior of an individual molecule.

[64] E. Wiedemann, "Zur Mechanik des Leuchtens," *Ann. der Physik*, 37 (1889): 177–248; pp. 236 ff.

He supposes the electrical equilibrium of the molecule to be disturbed and oscillations to be produced such that the two ends of some diameter become alternately charged with quantities of electricity $+e$ and $-e$. Then, following a certain number of oscillations, electrical equilibrium is re-established. Such a molecule represents "in some way" a Hertzian oscillator of very small dimensions. And its electrical oscillations are regarded as producing the spectral radiation.

It is interesting to note that from values of emitted radiation experimentally determined by Wiedemann, Ebert calculates that—"the amounts of free electricity which come into play in the molecular vibrations are therefore only the 1,500th part at most of those amounts we ought to suppose existing on the same quantity of matter in electrolytic phenomena."[65]

However, in a much longer article on the same subject published two years later in 1895, Ebert says that it is the "valency-charges" which oscillate and emit radiation.[66] Even a single atom which has only a single positive or negative charge can produce "valency-vibrations." In this case the atom is to be compared with a Hertzian oscillator connected on one side to earth.

Ebert advances his electromagnetic theory as an alternative to Wiedemann's dynamical theory. Both account equally for the quantities of radiation emitted. However, Ebert sees the former as avoiding difficulties which are encountered with the latter. In the first place he finds it difficult to accept the idea that an atom, which resists all chemical action, should have at the same time such a degree of elasticity that it can produce numerous and involved groups of spectral lines. This difficulty is especially great for monatomic metals such as mercury and cadmium. Furthermore, there is the difficulty with highly elastic atoms that if spectra are produced by their elastic deformations then the entire translational energy will become converted into vibrational energy and so lost as radiation. This is similar to the objection which Maxwell had raised against Boltzmann's view.

It would seem that both of these objections could also be made against Ebert's theory, especially the latter which is the more concrete. Ebert says that when two of his molecules collide the valency charges in each are displaced. Then during the free path the displaced charges oscillate and emit radiation. There would thus seem to be a similar process of over-all energy loss.

65 H. Ebert, "La mécanisme de la luminosité au point de vue de la théorie électromagnétique de la lumière," *Archives des Sciences Physiques et Naturelles,* 25 (1891): 489–503.
66 H. Ebert, "Electrische Schwingungen molecular Gebilde," *Ann. der Physik,* 49 (1893): 651–72; p. 653.

One objection which was actually made at the time against Ebert's theory involves the radiation emitted by a sodium atom. Ebert had estimated that this radiation, as determined by Wiedemann, was approximately the same as that emitted by a Hertzian oscillator of the size of an atom, charged with a "valency-charge" and vibrating with a frequency about that of the sodium D-lines. The objection was made by George Francis Fitzgerald at the 1893 meeting of the British Association for the Advancement of Science, and was reported as follows:

It was pointed out that the period of vibration of a simple oscillator of the size of an atom would be very many times more frequent than that of the sodium line, and that as the energy of radiation increases inversely as the cube of the wave-length it follows that the radiation of a simple Hertzian oscillator of the size of an atom might be many thousands of times what Wiedemann has found to be the radiating power of a sodium atom. It follows that sodium atoms must be complex Hertzian oscillators if they are Hertzian oscillators at all.[67]

A fellow-countryman of Fitzgerald's, Stoney, had already in 1891 dismissed the idea that spectral radiation is produced by a Hertzian discharge between charged molecules. Stoney advanced an electromagnetic theory of spectra by means of which he attempted to explain spectra of doublets and triplets such as those of sodium and magnesium respectively. But it was not only in this latter respect that he differed from Ebert. Stoney was the first to introduce orbiting electrons, defined as units of charge, as the seat of spectral radiation. However, this and his advocacy of the electromagnetic theory of spectra were the only positive aspects of his work. His explanation of doublets and triplets, considered in the following chapter, did not agree with observation and so was ignored.

It is interesting nevertheless to briefly consider Stoney's atomic molecular ideas. He proceeds to his electromagnetic theory of spectra, in a similar way to Ebert, by considering first a purely dynamical hypothesis and then converting this to an electrodynamic one. From the beginning we can see that he is thinking not surprisingly in terms of vortex atoms —"Let us then fix our attention on a particular molecule M, and suppose that a point P in it which acts on the aether has been set moving along some orbit within the molecule by the last of the inter-molecular encounters to which M has been subjected."[68]

[67] G. F. Fitzgerald, "Note on Professor Ebert's estimate of the radiating power of an atom, with remarks on vibrating systems giving special series of overtones like those given out by molecules," *Rep. British Assn.*, 61 (1893): 689–90.
[68] G. J. Stoney, "On the cause of double lines and of equidistant satellites in the spectra of gases," *Trans. Royal Dublin Soc. Sc.*, 4 (1888–92): 563–608; p. 569.

Stoney says we are ignorant about the forces under whose influence the point P continues its motion during the flight of the molecule. However, there is one case which allows mathematical treatment up to a certain point. This is when one or a few forces acting on P are predominant over all the others, and the treatment to be employed is the same as that employed in lunar and planetary theory. The true course of the point P is to be arrived at by first determining its dominant orbit, that is, its path under the dominant forces, and then by subjecting this orbit to perturbations.

In moving from the dynamical to the electromagnetic theory, Stoney reminds us that a charge of electricity is associated with each bond in the chemical atom. "There may accordingly be several such charges in one chemical atom, and there appear to be at least two in each atom."[69] Doubtless the two charges in each atom are of opposite sign— in order to have an electrically neutral atom. These charges, "which it will be convenient to call *electrons,* cannot be removed from the atom." This last is an important point. Now, says Stoney, if an electron is situated at a point P of the molecule, which undergoes the motion described above, the revolution of this charge will produce an electromagnetic wave in the surrounding ether.[70] Stoney's model is thus quite different from Ebert's, and one of the reasons must be that previously they had accepted different dynamical atomic/molecular theories of spectra.

Of the vortex atom theory of matter Stoney says that if it

be true, it is in the study of the dynamical, or rather kinematical, relations in, and in the neighborhood of, vortex rings and tangles, that we must put our hope. The vortex hypothesis, however, would suggest charges of magnetic movement rather than of statical electricity as associated with the atoms of ponderable matter. Perhaps both are present, and that the electrical charges are maintained by motions of the magnetism. Some motion of this kind must apparently be consequent on the velocity of over 30,000 metres per second with which the molecules in common with the rest of the earth, are travelling through the rectilinear vortices of the aether.[71]

But no matter how plausible Stoney tried to make his theory appear it was never taken seriously. Ebert's theory was similarly treated.

Nevertheless, there were many who agreed with Ebert and Stoney

69 *Ibid.,* p. 583.
70 The reader may object that I ought to have written "point P of the atom"; but Stoney does say "point P of the molecule." This apparent inconsistency is explained by assuming that Stoney, as many others, accepted the view that while individual atoms emit radiation, nevertheless each spectrum is produced by a molecule.
71 G. J. Stoney, "On the cause of double lines and of equidistant satellites in the spectra of gases," p. 606.

in principle, and in Europe there was a growing body of opinion in favor of the view that spectra had an electrical origin. But it was most difficult to know just how to proceed. Thus at the 1893 meeting of the British Association for the Advancement of Science, Oliver Lodge, Professor of Physics at University College in Liverpool, could do no more than state his belief that "radiation is due to the motion of electrified parts of molecules—not to the molecule as a whole."[72]

At the following year's meeting of the Association the old problem of reconciling kinetic theory with spectroscopic requirements was debated once more. And, not content with what they had contributed to the debate at the meeting, several of the participants wrote letters to *Nature* over the next several months. The first came from George Hartley Bryan, a Cambridge mathematician and an authority on kinetic theory:

> The objection which has been regarded by some as most antagonistic to the kinetic theory is that it does not afford an explanation of the spectra of gases. But is that required of it? If the luminosity of gases were due to the vibrations of the atoms within the molecules, certainly there *would* be a difficulty about regarding the molecules as rigid bodies; but then such a hypothesis would preclude a gas whose molecules were monatomic from having any optical properties whatever. To my mind, the electromagnetic theory of light entirely relieves the kinetic theory from the burden which has been imposed on it by its opponents, since if (for example) we regard the molecules as perfectly conducting hard spheres, spheroids, or other bodies moving about in a dielectric "vacuum" (i.e. space devoid of *ordinary* matter), we shall be able to account for the spectra by means of electromagnetic oscillations determined by surface-harmonics of different orders without interfering with the assumptions required for explaining the specific heats of gases.[73]

Another correspondent, S. H. Burbury, expressed confidence in this suggestion; but Edward P. Culverwell, a Fellow of Trinity College, Dublin, could not accept that the spectral motions would be independent of the other motions of the molecule.[74]

Fitzgerald also could not "follow" either Bryan's or what he understood to be Joseph Larmor's view, "that any help can be got by supposing spectral lines to be due to electromagnetic vibrations."[75] In reply to Fitzgerald, Schuster explained how with certain assumptions

[72] O. Lodge, "On the connection between aether and matter," *Rep. British Assn.*, 61 (1893): 688.

[73] G. H. Bryan, "Professor Boltzmann and the kinetic theory of gases," *Nature*, 51 (1894–95): 21.

[74] E. P. Culverwell, "The kinetic theory of gases," *Nature*, 51 (1894–95): 78–79; S. H. Burbury, "The ratio of the specific heats of gases, *Nature*, 51 (1894–95): 127.

[75] G. F. Fitzgerald, "The kinetic theory of gases," *Nature*, 51 (1894–95): 221–22.

one could relate a number of spectral rays to a "very restricted" number of degrees of freedom:

Most of us, I believe, now accept a definite atomic charge of electricity, and if each charge is imagined to be capable of moving along the surface of an atom, it would represent two degrees of freedom. If a molecule is capable of sending out a homogeneous vibration, it means that there must be a definite position of equilibrium of the "electron." If there are several such positions, the vibrations may take place in several such periods. Any one molecule may perform for a certain time a simple periodic oscillation about one position of equilibrium, and owing to some impact the electron may be knocked over into a new position. The vibrations under these circumstances would not be quite homogeneous, but if the electron oscillates about any one position sufficiently long to perform a few thousand oscillations, we should hardly notice the want of homogeneity. Each electron at a given time would only send out vibrations which in our instruments would appear as homogeneous. Each molecule would thus successively give rise to a number of spectral rays, and at any one time the electron in the different molecules would, by the law of probability, be distributed over all possible positions of equilibrium so that we should always see all the vibrations which any one molecule of the gas is capable of sending out. The probability of an electron oscillating about one of its positions of equilibrium need not be the same in all cases. Hence a line may be weak not because the vibration has a smaller amplitude, but because fewer molecules give rise to it. The fact that the vibrations of a gas are not quite homogeneous, is borne out by experiment. If impacts become less frequent by increased pressure, we should expect from the above views that the time during which an electron performs a certain oscillation is shortened; hence the line should widen, which is the case. I have spoken, for the sake of simplicity, as if an electron vibrating about one position of equilibrium would only do so in one period. If the forces called into play, by a displacement, depend on the direction of the displacement, there would be two possible frequencies. If the surface is nearly symmetrical, we should have double lines. The only weight I attach to these speculations lies in the illustration it affords that a number of spectral lines does not necessarily mean an equal number of degrees of freedom. In the existence of the "electron" I firmly believe; and this necessarily implies a very restricted number of variables.[76]

We recall how Stoney had attempted to overcome the difficulty of admitting as many independent motions in the molecule as there were lines in the spectrum. But his solution had required the spectral wavelengths to be related harmonically and none of the formulae found by Balmer, Rydberg, Kayser, and Runge met this requirement. Now Schuster suggests another possible explanation. Instead of assuming, as before, that a radiating molecule emits all the spectral wavelengths simultaneously, Schuster now assumes that the molecule emits only one or a few of the wavelengths at a given instant. However, all mol-

[76] A. Schuster, "The kinetic theory of gases," *Nature,* 51 (1894–95): 293.

ecules are simultaneously emitting different rays and so the entire spectrum is produced. Schuster's view is the one accepted today. Previously, however, credit has gone to Arthur William Conway for introducing it in 1907.[77]

The effect of Schuster's letter on Fitzgerald was apparently to remove the latter's skepticism concerning the electromagnetic view, even though Fitzgerald entirely missed Schuster's novel explanation of spectra. Fitzgerald actually says that to give spectral wavelengths Schuster analyzes molecular vibrational motions into their Fourier components.[78]

The last of the *Nature* letters is from Boltzmann who "cannot hope with Mr. Burbury, that Mr. Bryan will be able to deduce all of the phenomena of spectroscopy from the electromagnetic theory of light." Instead, Boltzmann proposes to "show it is possible to explain the spectra of gases while ascribing five degrees of freedom to the molecule, and without departing from Boscovich's standpoint."[79] So in spite of Maxwell's criticism Boltzmann had persisted in his view.

We saw above from Fitzgerald's first letter that Joseph Larmor, a lecturer in mathematics at Cambridge, was another early proponent of the electromagnetic view of spectra. Larmor is perhaps best remembered for his contributions to a dynamical theory of the ether. He published three major papers on this subject in the 1890s, and his book *Aether and Matter* appeared in 1900. Larmor's work, and particularly his first two papers which appeared in 1893 and 1895, is of much interest to us here because of the way in which he integrates the vortex atom theory of matter with electromagnetic theory.

In his first paper Larmor is concerned, among other things, with giving an explanation of magnetism for which he requires a circulating electric current.[80] His atoms are vortex atoms, whose cores are either vacuous or filled with a fluid devoid of elasticity, and the currents are represented by the vortex sheets on their surfaces. Larmor accepts the view that the atoms of the chemical elements are all combinations of a single type of "primordial atom." These "ultimate atoms" or "monads"

77 See Edmund Whittaker, *A History of the Theories of Aether and Electricity* (New York, 1960), 2: 106.

78 G. F. Fitzgerald, "On some considerations showing that Maxwell's theorem of equal partition of energy among the degrees of freedom of atoms is not inconsistent with the various internal movements exhibited by the spectra of gases," *Nature*, 51 (1894–95): 452–53.

79 L. Boltzmann, "On certain questions of the theory of gases," *Nature*, 51 (1894–95): 413–15.

80 J. Larmor, "A dynamical theory of the electric and luminiferous medium," *Phil. Trans.*, 185 (1894): 719–822; in J. Larmor, *Mathematical and Physical Papers* (Cambridge, 1929), I: 414–535; pp. 467–69.

must be taken "to be all quantitatively alike, except that some have positive electrification and others negative electrification."[81] This makes an interesting comparison with J. J. Thomson's earlier attempt to build the elements of the periodic table. The electrical view of matter has replaced the mechanical.

Larmor attributes spectral radiation to the vibrations of the atomic charges. The nature and frequencies of these vibrations and their associated radiations "depend only on the relative positions and motions of the vortex atoms in the molecule, and are quite unaffected, except indirectly, by irrotational motion (magnetic intensity) in the aether which they traverse."[82] Larmor refers to the growing experimental evidence that a gas cannot be made to produce a spectrum merely by heating it to a high temperature, "so that radiation in a gas must involve chemical action or, what is the same thing, electric discharge." Radiation will not be emitted

except in so far as the molecules are dissociated or their component atoms violently displaced with respect to each other. To allow the radiation to go on, such displacement must result on the whole in the performance of work against electric attractions, at the expense of the heat energy and chemical energy of the system, which must thus be transferred into electrical energy before it is radiated away.[83]

The above view of spectra underwent an important change when Larmor later attempted to overcome a difficulty encountered in his theory of magnetism. The remedy was "to suppose the core of the vortex ring to be made up of discrete electric nuclei or centres of radial twist in the medium."[84] These "monad charges—or electrons, as we may call them after Doctor Johnstone Stoney—must circulate very rapidly, in fact with velocities not many hundred times smaller than the velocity of radiation."[85] It is important for later considerations to note that in Larmor's view "the electron would be chemically an atom of very simple type such as e.g. that of hydrogen."[86] The electron, which seems now to be conceived of as a particle, has replaced the vortex ring as the ultimate atom of matter, although a unit composed of electrons has the appearance of a vortex ring as regards circulatory motion.

81 *Ibid.*, p. 475.
82 *Ibid.*, p. 486.
83 *Ibid.*
84 *Ibid.*, p. 515. This occurs in a section entitled "Introduction to free electrons," added to the paper after it had been presented to the Royal Society, but before it had been published.
85 *Ibid.*, p. 516.
86 *Ibid.*

Larmor was aware that it could be objected that a rapidly revolving system of electrons is effectively a vibrator, subject to intense radiation of its energy, and thus unstable. However, he demonstrates to his own satisfaction that radiation occurs only when the system has been disturbed, no radiation being emitted in the state of steady motion.

A system of two electrons of opposite sign revolving about one another would collapse, but this would not be so if the system "were roughly of the type of a ring of positive electrons revolving round an inner ring of negative ones."[87] This is a model intermediate between the mechanical-vortex and Bohr atoms. Larmor is confident that the origin of the numerous spectral lines of some "molecules" could be "illustrated" by such a system:

Thus in the analogous astronomical system of the Sun, Earth, and Moon, which has only nine coordinates, three for each body, there exist the much larger number of periodic inequalities or oscillations that are discussed in the Lunar Theory. The fact is that the oscillations of the coordinates of the system are not themselves harmonic or even exactly periodic; it is only when they are analyzed by the mathematical processes suitable to vibrations, or by a physical instrument such as a spectroscope which yields the same results, that the lines of the spectrum come into existence.[88]

Thus into the midst of all his new ideas Larmor throws an old and generally discarded idea of Stoney's.

However, in spite of any lingering mechanical ideas, the mood of spectroscopy had by 1895 become thoroughly 'modern.' The electrical view of spectra had become firmly implanted in the minds of spectroscopists. Even so, it was difficult to see ahead. As yet there was insufficient experimental data available to focus the attention of spectroscopists upon specific issues and so initiate fruitful scientific debate. With a few facts only, scientific speculation tends to run in too many different directions.

A few of the necessary facts were soon to be provided, however, and their discovery is the subject of the next section.

4. The Zeeman Effect

If the electromagnetic view of spectra required confirmation this was provided in 1896 by the discovery of the magnetic splitting of spectral lines, or Zeeman effect. It is interesting to note that its discoverer, the

---

[87] J. Larmor, "A dynamical theory of the electric and luminiferous medium—Part II: Theory of electrons," *Phil. Trans.*, 186 (1895): 695-743; in Larmor, *Mathematical and Physical Papers,* I: 597.
[88] *Ibid.*

Dutch experimental physicist Pieter Zeeman, had not been actively engaged in spectroscopic research. As with the Kundt–Warburg result and Hertz's demonstration that light is an electromagnetic radiation, this more recent event, which likewise was to greatly influence spectral theory, occurred outside the confines of spectroscopic research.

Writing in early 1897, Zeeman relates that it was "several years ago" while studying the Kerr effect that it occurred to him to investigate whether a magnetic field would in any way affect the light emitted by a flame.[89] In 1876 J. Kerr had shown experimentally that when plane-polarized light is regularly reflected from either pole of an iron electromagnet, the reflected ray has a component polarized in a plane at right angles to the ordinary reflected ray.[90] Thus Zeeman was not interested in spectroscopy as such, but in the effect of magnetism upon light.

Zeeman tells us how he attempted to explain to himself the way in which a magnetic field could affect the light of a flame:

If the hypothesis is true that in a magnetic field a rotatory motion of the aether is going on, the axis of rotation being in the direction of the magnetic forces (Kelvin and Maxwell), and if the radiation of light may be imagined as caused by the motion of the atoms, relative to the centre of mass of the molecule, revolving in all kinds of orbits, suppose for simplicity circles; then the period, or, what comes to the same, the time of describing the circumference of these circles, will be determined by the forces acting between the atoms, and then deviations of the period to both sides will occur through the influence of the perturbing forces between aether and atoms. The sign of the deviation of course will be determined by the direction of motion, as seen from along the lines of force. The deviation will be greater the nearer the plane of the circle approximates to a position perpendicular to the lines of force.[91]

As we have seen, the only theory of matter attributing the radiation of light to atoms revolving in a molecule was the vortex atom theory. This is what Zeeman has in mind here, and he relates it with another part of the teachings of Lord Kelvin (William Thomson) and Maxwell. The remainder of Zeeman's attempted explanation also shows the further influence of Kelvin on his thought.

It should be observed that Zeeman does not mention "electrons" or an electric charge of any kind in connection with his atoms and molecules. Thus at the time of his first thoughts on the subject he was thinking in terms of the mechanical, and not the electromagnetic, theory of vortex atoms.

---

[89] P. Zeeman, "On the influence of magnetism on the nature of the light emitted by a substance," *Phil. Mag.*, 43 (1897): 226–39.
[90] E. Whittaker, *A History of the Theories of Aether and Electricity*, I: 331.
[91] P. Zeeman, "On the influence of magnetism," pp. 230–31.

Zeeman tried unsuccessfully to influence the sodium spectrum by placing a sodium flame between the poles of a Ruhmkorff electromagnet. Some time later, in late 1894 or early 1895, he chanced to read in Maxwell's sketch of Faraday's life that in 1862 the latter had chosen the relation between magnetism and light as the subject of his very last experimental work. Faraday had also tried unsuccessfully to detect a change in the spectrum of a flame subjected to a "powerful" magnetic field. Zeeman's reaction to this was that if Faraday had thought of such a relation then perhaps it might yet be worth-while to try the experiment again. This time it was successful.

In Zeeman's experiment a piece of asbestos impregnated with common salt was placed in the flame of a Bunsen burner situated between the poles of an electromagnet. With the latter switched off the sodium D lines were seen as narrow, sharply defined lines on a dark background. When the magnet was switched on, however, the D lines became "distinctly widened." If a coal-gas flame fed with oxygen was used instead of the Bunsen burner the D lines became "perhaps three or four times their former width." Broadening was also observed with the red lithium lines.[92]

Zeeman was aware that increasing the pressure or temperature of a radiating gas also produces a broadening of its spectral lines, so he modified his experiment in order to eliminate these possible influences. The results with the magnetic field were the same as before.

If my understanding of Zeeman's paper is correct it was after he had been led, in part, by the vortex atom theory of spectra to seek a magnetic effect, and only after he had found such, that he turned to another theory of matter for its further explanation. Zeeman says: "A real explanation of the magnetic change of the period seemed to me to follow from Professor Lorentz's theory."[93] And he continues:

In this theory it is assumed that in all bodies small electrically charged particles with a definite mass are present, that all electric phenomena are dependent upon the configuration and motion of these "ions," and that light-vibrations are vibrations of these ions. Then the charge, configuration, and motion of the ions completely determine the state of the aether. The said ion, moving in a magnetic field, experiences mechanical forces of the kind above mentioned, and these must explain the variation of the period. Professor Lorentz, to whom I communicated these considerations, at once kindly informed me of the manner in which, according to his theory, the motion of an ion in a magnetic field is to be calculated, and pointed out to me that, if the explanation follow-

<hr />

92 *Ibid.*, pp. 226–27.
93 *Ibid.*, p. 231. See H. A. Lorentz, "La théorie électromagnétique de Maxwell et son application aux corps mouvants," *Archives Néerlandaises des Sciences Exactes*, 25 (1892): 363–552; *Versuche einer Theorie der elektrischen und optischen Erscheinungen in Bewegten Körpern* (Leiden, 1895).

ing from his theory be true, the edges of the lines of the spectrum ought to be circularly polarized. The amount of widening might then be used to determine the ratio between charge and mass, to be attributed in this theory to a particle giving out the vibrations of light.[94]

Suppose e is the charge and m the mass of a "vibrating ion" moving in the x–y plane in a magnetic field of uniform intensity H parallel to the z axis. The equations of relative motion of the ion are then

$$\left. \begin{aligned} m \frac{d^2x}{dt^2} &= -k^2x + \mathcal{L}H \frac{dy}{dt} \\ m \frac{d^2y}{dt^2} &= -k^2y - \mathcal{L}H \frac{dx}{dt} \end{aligned} \right\} \tag{1}$$

These are satisfied by

$$\left. \begin{aligned} x &= a\, \mathcal{L}^{sk} \\ y &= \beta \mathcal{L}^{st} \end{aligned} \right\} \tag{2}$$

provided that

$$\left. \begin{aligned} ms^2a &= -k^2a + eHs\beta \\ ms^2\beta &= -k^2\beta - eHsa \end{aligned} \right\} \tag{3}$$

where k is to be regarded as a known quantity.

The period of the motion T is of particular interest. If $H = 0$ it follows from (3) that

$$s = i\,\frac{k}{\sqrt{m}} = i\,\frac{2\pi}{T}$$

$$\text{or} \quad T = \frac{2\pi\sqrt{m}}{k}. \tag{4}$$

If H is not zero, it follows from (3) that approximately

$$s = i\,\frac{k}{\sqrt{m}}\left(1 \pm \frac{eH}{2k\sqrt{m}}\right).$$

Representing the period in this case by T', one has

$$T' = \frac{2\pi\sqrt{n}}{k}\left(1 \pm \frac{eH}{2k\sqrt{m}}\right). \tag{5}$$

Hence the ratio of the change of period to the original period is

$$\frac{eH}{2k\sqrt{m}} \quad \text{or} \quad \frac{e}{m}\frac{HT}{4\pi}. \tag{6}$$

A particular solution of equation (1) is that representing the motion of an ion in a circle. The period is given by (5)—the sign of the quantity $eH/2k\sqrt{m}$ depending upon the direction of motion, whether clockwise or anti-clockwise. The general solution of (1) shows that the ions also describe slowly rotating elliptical orbits. Thus if we imagine ourselves to be looking along the lines of force at the ions we see some of them moving in circles and emitting circularly-polarized light. Other ions appear to be stationary, but are really moving parallel to the direction of the magnetic field with unaltered period. Still others are moving in rotating elliptic orbits.

The motion of an ion moving in field free space may be resolved into a rectilinear harmonic motion parallel to the z axis and two circular motions (right-handed and left-handed) in the x–y plane. In a magnetic field parallel to the z axis the periods of the latter two motions become changed, while the period of the former remains unchanged. Thus the effect of a sufficiently strong magnetic field should be to cause a single spectral line to become a triplet—or at least one may expect the line to be wider and its edges to be formed of circularly-polarized light. This is what Lorentz predicted on the basis of his theory, and Zeeman now put it to the test of experiment.

Zeeman first viewed the sodium lines from a direction at right angles to that adopted in his first experiment, through holes bored in the poles of the electromagnet. A Nicol prism was positioned so that only the right-handed or left-handed circularly-polarized component was visible. If now the direction of the magnetic field is reversed the component ought to move—since the wavelength has been increased or decreased according to equation (5). This is what was observed to happen. When the phenomenon was viewed in the direction of the original experiment the edges of the widened line were found to be plane-polarized. In 1897 Zeeman used stronger magnetic fields and actually obtained doublet and triplet splittings of certain spectral lines. Thus at this date Lorentz's theory appeared to have been confirmed.

From the observed widening of the lines in his first experiments Zeeman calculated that the ratio of e/m was of the order of magnitude of 10^{-7}. He also deduced, incorrectly, that "the positive ions revolve, or at least describe the greater orbit," but shortly afterward he corrected himself and said that the negative ions revolve.[95]

With Zeeman's discovery the electromagnetic view of spectra became more firmly established, and we shall have an opportunity of appreciating this further in the epilogue. But now it is time to consider the attempts made prior to 1897 to give a mathematical theory of spectra.

95 P. Zeeman, "Doublets and triplets in the spectrum produced by external magnetic forces," *Phil. Mag.*, 44 (1897): 55–60, 255–59; p. 58.

V

||||||||||||||||||||||||||||||

Toward a Mathematical Theory of Spectra

A CONSIDERATION of the attempts made to give a mathematical theory of spectra—that is, one which would wed spectral determinations, and in particular spectral series formulae, to an atomic or molecular model—affords a true measure of the extent, or limit, of the nineteenth century's understanding of spectra.

Two courses were open to the theorist. On the one hand, he could take an existing atomic or molecular model, or some adaptation of it, and attempt mathematically to derive the empirical laws. On the other hand, he could seek an instance of some elastic body vibrating according to a law identical in mathematical form to one of the series formulae, and having found it, then attempt to arrive by analogy at the physical nature of atoms and molecules. These alternative approaches may respectively be called the synthetical and the analytical.

To begin with the synthetical, what would appear to be the first attempt in this category was made in 1869 by Lecoq de Boisbaudran, in connection with his work discussed above in chapter III.[1] As with most of the attempts to be considered, it would serve no useful purpose to give a full account of Lecoq's speculations. For in addition to exceeding reasonable limits imposed by contemporary science, and perhaps for this reason, they were not adopted by his contemporaries. Thus a brief consideration will suffice.

Of all the motions which a "molecule" may have, Lecoq takes that of rotation about itself to be "the most simple and most inevitable." And thus provided with a periodic motion he continues:

[1] See p. 106.

If the molecule (the axis and the centre of gravity being supposed immovable) is a solid of revolution about its axis, no vibration will be communicated to the surrounding aether; but if the molecule has inequalities, each time that one of these inequalities passes by one of the meridian planes, supposed fixed in space, a wave will be produced whose axis will be contained in the fixed plane considered. The distances (measured in longitude) between the several inequalities, will determine the time passed between the emission of one wave and the next; whence the wave-lengths. On the number of inequalities and of their relative positions, will depend the formation of a first spectrum which will be the group of rays *characteristic of the molecule,* the unit, the element, of which the several parts of the entire spectrum will be formed. If the molecule possesses only a simple motion of rotation, its spectrum will therefore be reduced to a single group of lines, having no analogies in other parts of the luminous scale.[2]

These are the first of the basic suppositions of Lecoq's theory. However, we already know from chapter II that by 1869 the molecular theory of spectra, which took account of kinetic theory considerations, was generally accepted. And according to it, spectra have their origin only in vibrational and never in rotational motions. Consequently it is clear why Lecoq's "fantastic world of atoms," as Rydberg later referred to it, was ignored.[3]

Stoney, we recall, was one of the originators of the molecular theory of spectra, and it was he who next attempted to give a mathematical theory of spectra. This we considered in chapter III.[4] Although Stoney employs an acoustical analogy it is not in the analytical sense spoken of at the beginning of this chapter. He does not attempt to reason back by analogy to the nature of the molecules. What he does is to postulate that the radiation emitted by a molecule becomes separated into its various components in traversing the prism of a spectroscope. In this way he attempts to fuse a vague molecular model with observation by stipulating the law of distribution of spectral lines.

Schuster was later to demonstrate that spectral lines are not distributed according to a simple harmonic law, as Stoney had believed.[5] And later still Balmer found the true formula for the hydrogen series.[6]

Meanwhile, the British had adopted the vortex atom theory of matter and it may resonably be assumed that attempts were made to derive the hydrogen series formula from the mechanical properties of vortex atoms. But in this no one succeeded. Likewise no success was had with

[2] F. Lecoq de Boisbaudran, "Sur la constitution des spectres lumineux," *Comptes Rendus,* 69 (1869): 445–51; p. 447.

[3] J. R. Rydberg, "Recherches sur la constitution des spectres d'émission des éléments chimiques," *Konliga Svenska Vetenskaps-Akademiens Handligar,* 23 (1888–89), No. 11, p. 6.

[4] See p. 110.

[5] See p. 122.

[6] See p. 131.

the German atom. Here we know of one attempt—already mentioned in chapter IV. Treating the ether envelope of an atom as a hollow sphere, J. Loschmidt investigated its vibrations in the hope of arriving at spectral series formulae.

By the time the next synthetical attempts were made the electromagnetic was displacing the elastic theory of radiation. Perhaps the earliest attempt to calculate spectral wavelengths on the basis of the former theory was that made in 1887, in a paper on the dispersion of light.[7] The author was Franz Koláček, a Brünn schoolteacher. Starting from Maxwell's general equations of the electromagnetic field and applying them, with the necessary modifications, to the case of the electromagnetic vibrations of a sphere, Koláček arrives at an equation yielding the wavelengths emitted by a gas composed of spherical molecules. This, he says, should be applicable to monatomic mercury vapor, with the qualification that, as he has discussed only the simplest kind of vibrations the equation will not yield all of the mercury wavelengths. However, he does not compare the calculated with the observed wavelengths. Hermann Ebert later did so and found no agreement.[8] Koláček merely adds that it would be interesting to consider other molecular types in order to find a structure corresponding to the recently discovered Balmer formula.

But if Koláček did consider other molecular types in regard to the hydrogen series formula he never met with any success. In fact no attempt within the synthetical category was successful in producing in the nineteenth century either a mechanical or an electromagnetic theory yielding Balmer's series, let alone those of Rydberg and Kayser and Runge. Attention consequently became focussed on what appeared to be a more simple problem—that of explaining the fact, first noted by Hartley, that in many doublet series the separation of the components of a doublet is a constant for the series. It was in relation to this problem that Koláček published a lengthy paper some eight years later in 1896.

Koláček considers the electromagnetic oscillations of a conducting sphere which is capable of being polarized and which is immersed in the "dielectric polarizable aether."[9] And he succeeds in demonstrating that at least two series of double lines must exist, the separations of

7 F. Koláček, "Versuch einer Dispersionserklarung vom Standpunkte der electromagnetischen Lichttheorie," Ann. der Physik, 32 (1887): 224–55; pp. 237–39.

8 H. Ebert, "La mécanisme de la luminosité au point de vue de la théorie électromagnétique de la lumière," Archives des Sciences Physiques et Naturelles, 25 (1891): 489–503; p. 493.

9 F. Koláček, "Ueber electrische Oscillationen in einer leitenden und polarisationsfähigen Kugel. Ein Beitrag zur Theorie der Spectra einfachster Beschaffenheit," Ann. der Physik, 58 (1896): 271–310.

whose doublet components either remain constant or decrease, thus corresponding respectively to the subordinate and principal series of the alkalis.

Koláček was apparently greatly influenced by the work of Kayser and Runge, and in keeping with their views he believes that the principal series might be produced by a molecule different from that producing the subordinate series. If this is the case the radius of the latter molecule, he says, must be 1.61 times that of the former.[10] "The principal series corresponds then to the oscillations of the dissociated components (perhaps the atoms), the subordinate series to those of the non-dissociated group."[11]

As far as I have been able to determine, Koláček's work received little attention, a fate which, as will be seen, was also that of the other work on doublets. Perhaps most spectroscopists were like Runge, who thought that for a theory to be at all plausible it must explain more than one of the number of "regularities" to be found in spectra.[12] But then Runge was unable to do any better himself.

One objection to Koláček's work was that it predicted infinite series of doublets, whereas Kayser and others were convinced that spectral series had finite limits.[13] Koláček had attempted to overcome this difficulty by saying that as the series proceeded to infinity the intensity of the lines became infinitesimal and so were invisible.

Meanwhile, in the interval between Koláček's two papers, Stoney had advanced a different explanation of doublets.[14] Assuming the motions of the two units of charge (or electrons) within a vortex atom to be elliptical, Stoney considers the nature of the radiation emitted by one of these when its motion is subjected to various perturbations. The treatment employed is "the same as that with which we are familiar in the lunar and planetary theories."[15]

When the electron's motion suffers no perturbation the electron emits, to follow Stoney's acoustical analogy, a certain note. As in Stoney's earlier theory, discussed above in chapter III, this becomes separated into its various partials, given by a Fourier series, when the radiation traverses a prism. If, now, there is a perturbation of the elec-

[10] Kayser gets this wrong, confusing mass with radius. *Handbuch der Spectroscopie* (Leipzig, 1902), 2: 606.
[11] F. Koláček, "Ueber electrische Oscillationen in einer leitenden und polarisationsfähigen Kugel," p. 305.
[12] C. Runge, "On the line spectra of the elements," *Nature,* 45 (1891–92): 607–8; p. 607.
[13] H. Kayser, *Handbuch,* 2: 607.
[14] See p. 188.
[15] G. J. Stoney, "On the cause of double lines and of equidistant satellites in the spectra of gases," *Trans. Royal Dublin Soc. Sc.,* 4 (1888–92): 563–608; p. 569.

tron motion in the form, say, of an apsidal motion of the ellipse in its own plane, Stoney demonstrates that two notes of different intensity are emitted instead of one. When each of these is resolved into its partials the result is a series of doublets in which the separation of the components of a doublet is a constant for the series. The less refrangible component of each doublet will be more or less intense than the more refrangible component, depending upon the direction of the apsidal motion with respect to the elliptical motion. In addition, Stoney demonstrates that other perturbations may, for example, give rise to triplets or cause the displacement of single lines.

Stoney's main interest is, however, in doublets, and he applies his theory to the series of sodium doublets given by Rydberg. From their appearance—that is, the separation and relative intensities of the doublet components—he is able to reason back to the nature of the motion of an electron within a sodium molecule.[16] Thus he finds, for example, that the motion of the electron producing the sodium D lines is on a "long-shaped" ellipse which it traverses 1,894 times while the ellipse moves round once in the same direction. This makes an interesting contrast with William Thomson's earlier explanation of the production of the D lines by a mechanical vortex atom.

Thus Stoney hoped to be able to make precise determinations of the motions of electrons within molecules. However, as he himself acknowledges, his theory meets in most cases with a serious difficulty when two or more perturbations have to be combined—for the phase of one perturbation relative to another cannot be known. And thus the actual motion of an electron can never be completely determined.

No one before had questioned the postulate of Stoney's earlier theory, repeated here in his electromagnetic theory, that radiation emitted by a molecule is resolved by a prism into its various components according to a Fourier series. Schuster had of course demonstrated that spectral lines are not distributed according to a simple harmonic law. But Runge, in 1892, claimed that such a resolution was impossible and he and Stoney debated the issue, obscurely and unsatisfactorily, in a series of letters to *Nature*.[17] It is curious that Runge did not simply object that, as he very well knew, no series of doublets formed a Fourier series. Kayser did raise this objection, but only later. Others may have seen it from the beginning. In any case Stoney's work received little attention.

16 See footnote 70 of chapter IV.
17 C. Runge, "On the line spectra of the elements," *Nature*, 45 (1891–92): 607–8; p. 607, and *ibid.*, 46 (1892): 100, 200, and 247. Stoney, *Nature*, 46 (1892): 29, 126, 222, and 268.

We turn now to what I have called analytical attempts, and begin with Deslandres. It is recalled that in 1887 he had shown the distribution of lines in band spectra to be given by the function

$$Am^2 + Bn^2 + \sqrt{Cp^2 + \gamma}$$

where the variables m, n, and p, each takes a series of integral values and A, B, C, and γ, are constants. To Deslandres this suggested a "very clear analogy" between the distribution of spectral lines and the succession of notes emitted by a vibrating solid body.[18] For the vibrations of a rectangular prism are given by

$$K\sqrt{\frac{m^2}{a} + \frac{n^2}{b} + \frac{p^2}{c}}$$

where m, n, and p take whole number values and a, b, and c are the dimensions of the plate.

The analogy is therefore manifest and, when this function of three parameters will have been determined experimentally for these spectra, it will evidently be curious to take up the problem in the reverse sense and to investigate the conditions at the surfaces which would give an identical distribution of sounds. The question is, in fact, arrived at a point where the help of mathematical analysis and of the theory of numbers can be very useful.[19]

But while the analogy continued to interest Deslandres he never progressed beyond saying that if "one admits the atomic hypothesis, the atoms vibrate as small bodies having three dimensions and a determined form."[20]

To digress briefly on Deslandres's remark, I have found no spectroscopist who did not admit an atomic theory. As we have seen, "understanding spectra" became almost synonymous with "understanding atoms and molecules," and to spectroscopists atoms and molecules were as real as spectra themselves. Thus the end of the century's anti-atomic movement, which had its origin in the application of thermodynamics to chemical phenomena, could only have met with resistance from spectroscopists.[21] It had apparently no influence on spectroscopy.

The next analytical attempt is a most interesting one. It was pub-

18 H. Deslandres, "Loi de rèpartition des raies et des bandes, commune à plusieurs spectres des bandes. Analogie avec la loi de succession des sons d'un corps solide," *Comptes Rendus*, 104 (1887): 972–76; p. 976, fn.
19 *Ibid.*
20 H. Deslandres, "Proprieté fondamentale commune aux deux classes des spectres. Caracteres distinctifs de chacune des classes. Variations périodique à trois paramètres," *Comptes Rendus*, 110 (1890): 748–50; p. 750, fn.
21 On the anti-atomic movement and its origins see J. T. Merz, *A History of European Thought in the Nineteenth Century* (London, 1904–12), 2: 183–87.

lished by Alexander Herschel in 1896 in reply to the charge that while spectral series formulae were known, yet "no approach to a physical explanation of the series has yet been arrived at."[22] Herschel thought this "hardly correct," at least in the case of the hydrogen series.

Herschel had discovered that for an open pipe to sound its mth harmonic when traversed by air traveling at a speed $(n/m) \cdot v$ (where v is the velocity of sound in air and m and n are integers with $n < m$), the pipe has to be shortened by a fraction $(n/m)^2$ of its length. If the length of the pipe is kept unaltered, however, air of the same velocity will cause the pipe to emit a note whose frequency is to that of the original harmonic as $(1 - n^2/m^2)$: 1. Thus if the latter frequency is A the new frequency is given by $1/\lambda = (1 - n^2/m^2)A$. And if one places $n = 2$ one has an equation similar to Balmer's formula $(1/\lambda = (1 - 2^2/m^2)A)$.

Herschel, however, does not compare the hydrogen atom to an open pipe. Rather, he attempts to adapt his result to the prevailing vortex atom theory of matter. He imagines a flute with open ends bent into a circle so that its two ends meet. When air passes around the tube the harmonics produced are identical to those produced by a straight open tube half as long—since the circular tube's two halves behave simply as two consecutive open tubes.

A circular revolving ring of fluid, rather than a stream of fluid through a straight tube, thrown into resonance, and emitting vibrations from a lateral point of its circumference, seems thus to be the true construction of a vibrating atom which conforms perfectly to Balmer's law; and the three series of lines found (in certain spectra) would thus seem to indicate three ring-clusters interlacing, among whose circuits, of all possible velocities, only those speeds are perceivable by the spectroscope which fulfill the conditions of resonance in their circuits in 1, 2, 3, or more stationary wave-length undulations along them.

Now that an electro-magnetic view is taken of the nature of light and etherial radiations, the annular vortices also now generally regarded as composing atoms' structures may, perhaps, not inappropriately, be likened to grammerings endowed with annular as well as with vortical rotations.[23]

As Herschel himself recognized, however, his acoustical example merely relates two frequencies or wavelengths to one another, whereas Balmer's law relates a series of wavelengths to a "fundamental" wavelength. Nevertheless, Herschel felt that his analogy would "probably furnish us in some form with a real explanation of the law."[24]

Somewhat earlier in the decade Fitzgerald had obtained close approx-

---

22 A. S. Herschel, "On a relation between the spectrum of hydrogen and acoustics," *Observatory*, 19 (1896): 232–34; p. 232.
23 *Ibid.*, pp. 233–34. A gramme ring is an armature having a ring winding.
24 *Ibid.*, p. 232.

imations to the hydrogen and alkali series formulae in considering the vibrations of an open chain of magnets.[25] Of the two final contributors to be considered in the analytical category, the second, Lord Rayleigh, refers directly to this work, while the first, Schuster, would seem to have had it in mind when late in 1896 he wrote: "We may imagine a number of particles in a row, and raise the question whether we can imagine some connection between them such that they should be able to vibrate in a series of periods similar to those observed in the spectra of gases."[26] While to Schuster it did not seem impossible to imagine some connection such that for the lower frequencies the connecting forces regulate the frequency, while for the higher frequencies the frequency tends to become equal to that of the separate particles, he nevertheless proposed nothing specific. All he said, without offering any explanation, was that looked "at from a different point of view we may say that, if we could imagine a rod having elastic properties such that the relation between the velocity of a wave along it and the wavelength is $V = a\lambda - b\lambda^3$ it would, if vibrating freely, give out a number of notes, the relative frequencies of which would be the same as that of the luminous vibrations given out by a hydrogen molecule."[27]

But in the following year Lord Rayleigh called attention to "one circumstance which suggests doubts whether the analogue of radiating bodies is to be sought at all in ordinary mechanical or acoustical systems vibrating about equilibrium."[28] For the latter give rise to equations involving the square of the frequency, whereas spectral series formulae involve only the first power of the frequency. "For example, this is the case with Balmer's formula. Again, when the spectrum of a body shows several doublets, the intervals between the components correspond closely to a constant difference of frequency, and could not be simply expressed in terms of squares of frequency. Further, the remarkable law, discovered independently by Rydberg and Schuster, connecting the convergence frequencies of different series belonging to the same substance, points in the same direction."[29] Rayleigh concludes by saying that the first power of the frequency suggests kinematical rather than dynamical considerations.

Thus Rayleigh tended to bring to an end the entire analytical approach and so to discard the acoustical analogy which had been so

[25] This is mentioned by Lord Rayleigh, "On the propagation of waves along connected systems of similar bodies," *Phil. Mag.*, 44 (1897): 356–62; p. 361.
[26] A. Schuster, "On a new law connecting the periods of molecular vibrations," *Nature*, 55 (1896–97): 200–1; p. 201.
[27] *Ibid.*
[28] Lord Rayleigh, "On the propagation of waves," p. 362.
[29] *Ibid.*

much a part of the understanding of spectra in the nineteenth century. And as synthetical attempts to give a mathematical theory of spectra had also been unsuccessful no one knew quite what to do. For the synthetical and analytical approaches had seemed the only possible ways of getting to the nature of atoms and/or molecules, depending on one's point of view, and thus to a satisfactory understanding of spectra. But by early 1897 it appeared that, in spite of the earlier optimism of Rydbergs and Runges, this goal might not be attained in the near future, if indeed ever.

In the very same year, however, a third alternative way was unexpectedly opened up. Its nature and how it was made possible are the subjects of the Epilogue.

||||||||||||||||||||||||||||||||

EPILOGUE—
The Corpuscle: Beginning and End

Pieter Zeeman's first paper on the magnetic broadening of spectral lines, considered in chapter IV, was published in the *Philosophical Magazine* in March, 1897. On the last day of the following month J. J. Thomson announced the discovery of the corpuscle, or the particle later to be called the electron, in a lecture at the Royal Institution in London. Because of this timing, and also because Zeeman's negative ion turned out to be identical with Thomson's corpuscle, some writers have suggested that Thomson's discovery was influenced by Zeeman's earlier discovery.[1] This is a view for which, as I shall show, a very strong case can be made on grounds perhaps never before considered.

We must begin by briefly considering the state of the debate on the nature of the cathode rays prior to their identification as corpuscles or electrons. Our starting point is J. J. Thomson's Presidential Address to the Mathematics and Physical Science Section of the British Association for the Advancement of Science, in September, 1896. At that date, and as had been the case for some time, two views were held as to the nature of the cathode rays. As with theories of matter, we also find here a national difference. The British view was, in Thomson's words, "that they are particles of gas carrying charges of negative electricity, and moving with great velocities which they have acquired as they travelled through the intense electric field which exists in the neigh-

---

[1] See G. E. Owen, "Discovery of the electron," *Annals of Sc.*, 11 (1955): 173–82. Owen's view is supported by C. L. Maier in his doctoral dissertation: "The role of spectroscopy in the acceptance of an internally structured atom, 1860–1920" (University of Wisconsin, 1963–64), pp. 323–31.

bourhood of the negative electrode."[2] The German view, on the other hand, was that the cathode rays are ether waves. Each side could support its view with convincing arguments and at the same time attack the opposing view with equally convincing arguments.

There was one aspect of the cathode rays, however, which neither side could explain satisfactorily. In 1893 Philip Lenard, the leading advocate of the German view, had used a discharge tube provided with a thin aluminum window and had found that a phosphorescent screen placed outside the window in direct line with the cathode rays became luminous. Likewise, photographic plates were affected and electrified bodies became discharged. Finally, he found that the rays outside, as those inside, the tube were deflected by a magnet. Lenard therefore concluded that the rays outside the tube (Lenard rays) were cathode rays which had passed through the aluminum window. The fact that the cathode rays could apparently pass through matter provided the strongest grounds for the ether view.

J. J. Thomson, the leading proponent of the British view, explained the Lenard rays by saying that the air or gas outside the aluminum window must be traversed by the recently discovered Röntgen rays. These would ionize the gas or air molecules and the resulting ions would be set in motion by the electrostatic charge on the aluminum window. Hence, the Lenard rays as the cathode rays would be charged molecules. This explanation, however, was not fully satisfactory, and Thomson knew it. But all he said by way of conclusion at the British Association Meeting was: "Though there are some points in the behaviour of these Lenard rays which do not admit of a very ready explanation from this point of view, yet the difficulties in its way seem to me considerably less than that of supposing that a wave in the aether can change its velocity when moving from point to point in a uniform field."[3]

One difficulty with Thomson's explanation of the Lenard rays was that these latter could travel a distance of half a centimeter through air at atmospheric pressure before the luminosity produced by them in a phosphorescent screen fell to half its maximum value.

Now the mean free path of the molecule of air at this pressure is about 10^{-5} centimetres, and if a molecule of air were projected it would lose half of its momentum in a space comparable with the mean free path. Even if we suppose that it is not the same molecule that is carried, the effect of the obliquity of the

2 J. J. Thomson, *Rep. British Assn.*, 64 (1896): 699–706; p. 701.
3 *Ibid.*, p. 702.

collisions would reduce the momentum to one-half in a short multiple of that path.[4]

Thus Thomson was in a dilemma. He could not explain the large path lengths of the Lenard rays; and while Lenard could easily do so, Thomson could not accept the German view that ether waves are bent in a magnetic field. There seemed to be no way out.

One solution would have been to explain the path lengths by assuming the Lenard rays to consist of very small particles. However, such a particle would have to have dimensions very much smaller than an atom, and to a disciple of the vortex atom, or any other atomic theory in 1896, the idea of such a free particle was unthinkable. Even if Thomson had gone along with Larmor's development of the vortex atom theory, and we do not know that he did not, the smallest charged particle available—Larmor's electron—had nevertheless the dimensions of a hydrogen atom.

This, then, was Thomson's position with regard to the cathode and Lenard rays in the fall of 1896, when Zeeman was announcing his results in Amsterdam.

Hitherto, no one seems to have called attention to the fact that notice of Zeeman's discovery appeared in the Christmas Eve, 1896, edition of *Nature*.[5] This came to the attention of Larmor who was then away in the country in Ireland. Larmor, as he tells us, was cognizant of the results to be obtained by subjecting a pair of mutually orbiting electrons to a magnetic field—"Taking the masses of the ions to be comparable with that of a hydrogen atom, the spectral effect would be inappreciable." Larmor pointed out the circumstance to Professor Lodge, and suggested the importance of confirming Zeeman's experiment, which Lodge soon succeeded in doing.[6]

Thus, from Zeeman's results, the vortex atom theory of matter was seen to be in error. This consideration must quickly have come to the attention of Thomson, a colleague of Larmor's at Cambridge, then living with the Lenard-ray dilemma. It was easy to see, however difficult it might be to accept—for it would entail abandoning the vortex atom theory of matter—that, in compliance with Zeeman's result, to admit the existence of a free charged particle of mass very much smaller than that of a hydrogen atom, would be to afford an explanation of the relatively large Lenard-ray path lengths. It was in this way, I sug-

[4] J. J. Thomson, "Cathode Rays," *Proc. Royal Inst.,* 15 (1897): 419–32; p. 430.
[5] Report of October 31, 1896, meeting of Royal Academy of Sciences, Amsterdam, in *Nature,* 55 (1896–97): 192.
[6] J. Larmor, *Mathematical and Physical Papers,* 2: 140.

gest, that J. J. Thomson was led, in late 1896 or early 1897, to a reexamination of the properties of cathode rays. He had a definite goal in mind—the discovery of what for an adherent of the vortex atom theory of matter was a truly strange entity—a free subatomic particle.

At a meeting of the Cambridge Philosophical Society on February 8, 1897, Thomson gave the results of experiments made on the charge carried by the cathode rays and on their magnetic deflection. The charge was negative, and most important, the "magnetic deflection of the cathode rays in air, hydrogen, carbonic acid gas and methyl iodide is the same provided the mean potential difference between the anode and the cathode is the same."[7]

Almost two months passed and then on April 30 Thomson announced the discovery of the corpuscle. At the very end of his address he remarked that it "is interesting to note that the value of e/m which we have found from the cathode rays, is of the same order as the value 10^{-7} deduced by Zeeman from his experiments on the effect of a magnetic field on the period of the sodium light."[8]

Nevertheless, the corpuscle or electron was not an entity which could be readily accepted. As Thomson himself wrote, long after the exciting event:

At first there were very few who believed in the existence of these bodies smaller than atoms. I was even told long afterwards by a distinguished physicist who had been present at my lecture at the Royal Institution that he thought I had been "pulling their legs." I was not surprised at this, as I had myself come to this explanation of my experiments with great reluctance, and it was only after I was convinced that the experiment left no escape from it that I published my belief in the existence of bodies smaller than atoms.[9]

When he first published this belief Thomson drew on Lockyer for support, saying that the latter had advanced "weighty arguments, founded on spectroscopic considerations, in favour of the composite nature of the elements."[10] Moreover, Lockyer and Thomson now found themselves in the same position in that each claimed to have demonstrated the widely held view that the elements are compound entities.

It is at first sight paradoxical that in spite of this general belief Thomson's corpuscle was slow to be accepted. But the consideration of Lockyer's dissociation hypothesis in chapter II enables us to suggest what are perhaps two new reasons for this. First, and perhaps more impor-

---

7 J. J. Thomson, "On the Cathode Rays" [1897], *Proc. Cambridge Phil. Soc.,* 9 (1898): 243–44.
8 J. J. Thomson, "Cathode Rays," p. 432.
9 Sir J. J. Thomson, *Recollections and Reflections* (New York, 1937), p. 341.
10 J. J. Thomson, "Cathode Rays," p. 432.

tant, is the attitude met with in Hartley's criticism of the hypothesis. Hartley found it inconceivable that the complex iron "molecule" could consist of as many as 1,200 parts. But now Thomson's discovery entailed that the simplest "molecule" of all, the hydrogen chemical atom, consists of half as many parts again! The nineteenth century generally was not prepared for that.[11] Second, the widespread though not complete rejection of Lockyer's hypothesis had made a terrestrial dissociation of the elements seem unlikely.

For spectroscopy the discovery and acceptance of the electron was, like the Kundt–Warburg result in Britain, both an end and a beginning. The vortex atom theory of matter, with its indestructible atoms, was itself destroyed.[12] The conclusion to which attempts at a mathematical theory of spectra were pointing, namely, that existing atomic models were quite inadequate, was confirmed. But at the same time a fresh route to a mathematical theory was promised. The nature of the problem though was changed. Now it seemed that the materials, so to speak, of atoms were known and the problem was to put these together so as to obtain the spectral series formulae. Hope was rekindled, but on the threshold of modern physics the problem of giving a mathematical theory of spectra seemed as formidable as before.

---

[11] I say "generally" because some like Kayser (see above p. 99), perhaps took readily enough to the idea.

[12] Lord Kelvin (William Thomson) had been assailed by doubts about the vortex atom theory from about 1886. In 1887 he came to the conclusion that a vortex ring is "essentially unstable and that its fate must be to become dissipated." But it was not until 1898 that he abandoned the vortex atom saying: "I am afraid that it is not possible to explain all the properties of matter on the vortex-atom theory alone." See S. P. Thompson, *Life of Lord Kelvin* (London, 1910), 2: 1046–47.

Bibliography

The history of nineteenth-century spectroscopy has as yet received little attention. In considering the subject one can hardly do better than start with Heinrich Kayser's *Handbuch der Spectroscopie,* 8 vols. (Leipzig, 1901–1912). It would seem that Kayser has missed no relevant source. Yet only in the first 120 pages of the first volume does he attempt to write a history of nineteenth-century spectroscopy. And this is a disappointing effort, particularly for the period from 1860 to 1900. Much the same can be said of the historical part of E. C. C. Baly's *Spectroscopy,* 2 vols. (3d ed.; London, 1929).

An account of certain aspects of nineteenth-century spectroscopy is also to be found in Ferdinand Rosenberger's *Geschichte der Physik,* 3 vols. (Braunschweig, 1887–1890), yet at times this is superficial.

The subject of Clifford L. Maier's doctoral dissertation—"The Role of Spectroscopy in the Acceptance of an Internally Structured Atom, 1860–1920" (Univ. of Wisconsin, 1963–1964)—overlaps in part with that of the present volume, but our respective treatments of the common area are quite different.

Secondary Works

Books

Anderson, David L. *The Discovery of the Electron.* Princeton, 1964.
Baly, E. C. C. *Spectroscopy.* 3d ed.; 2 vols. London, 1929.
Brush, Stephen G. *Kinetic Theory.* 2 vols. London, 1966.
Campbell, L., and Garnett, W. *Life of James Clerk Maxwell.* London, 1884.
Clerke, Agnes. *A Popular History of Astronomy during the Nineteenth Century.* 4th ed. London, 1902.

D'Albe, E. E. Fournier. *The Life of Sir William Crookes O.M., F.R.S.* New York, 1924.

Kayser, Heinrich. *Handbuch der Spectroscopie.* 8 vols. Leipzig, 1901–12.

Knight, David M. *Atoms and Elements.* London, 1967.

Lindroth, Sten, ed. *Swedish Men of Science, 1650–1950.* Stockholm, 1952.

Maier, Clifford Lawrence. "The Role of Spectroscopy in the Acceptance of an Internally Structured Atom, 1860–1920." Ph.D. diss., University of Wisconsin, 1963–64.

Merz, J. T. *A History of European Thought in the Nineteenth Century.* 4 vols. London, 1904–12.

Rosenberger, F. *Die Geschichte der Physik.* 3 vols. Braunschweig, 1887–90.

Thompson, S. P. *Life of Lord Kelvin.* 2 vols. London, 1910.

Whittaker, E. *A History of the Theories of Aether and Electricity.* 2 vols. New York, 1960.

PAPERS

Daub, Edward E. Atomism and thermodynamics. *Isis* 58 (1967): 293–303.

Farrar, W. V. Nineteenth-century speculations on the complexity of the chemical elements. *British J. Hist. Sc.* 2 (1965): 297–323.

Finn, Bernard S. Thomson's dilemma. *Physics Today* 20 (1967), no. 9: 54–59.

Hamor, W. A. David Alter and the discovery of chemical analysis. *Isis* 22 (1934–35): 507–10.

Kedrov, B. M. Toward a methodological analysis of scientific discoveries. *Soviet Studies in Philosophy* 1 (1962–63), no. 1: 45–57.

Owen, G. E. Discovery of the electron. *Annals of Sc.* 11 (1955): 173–82.

Silliman, Robert H. William Thomson: Smoke rings and nineteenth-century atomism. *Isis* 54 (1963): 461–74.

Sources

BOOKS

Ames, J. S. *Prismatic and Diffraction Spectra, Memoirs by Joseph von Fraunhofer.* New York and London, 1898.

Ångstrom, A. J. *Recherches sur le Spectre Solaire.* Uppsala, 1868.

Becquerel, A. C. *Traité Experimentale de L'Électricité.* Paris, 1836.

Foucault, L. *Recueil des Travaux Scientifiques de Léon Foucault.* Paris, 1878.

Hasenöhrl, F., ed. *Wissenschaftliche Abhandlungen von Ludwig Boltzmann.* 3 vols. Leipzig, 1909.

Herschel, J. W. F. *A Treatise on Astronomy.* Philadelphia, 1834.

Kayser, H. *Lehrbuch der Spectroscopie.* Berlin, 1883.

Kirchhoff, G. *Researches on the Solar Spectrum and the Spectra of the Chemical Elements.* Translated by H. E. Roscoe. 2 vols. London, 1862–63.

Larmor, J. *Mathematical and Physical Papers.* 2 vols. Cambridge, 1929.

——— ed. *Memoir and Scientific Correspondence of Sir George Gabriel Stokes.* 2 vols. Cambridge, 1907.

———. *Mathematical and Physical Papers of Sir George Gabriel Stokes.* 5 vols. Cambridge, 1880–1905.

———. *Mathematical and Physical Papers by the Right Honourable Sir William Thomson.* 6 vols. Cambridge, 1882–1911.

Liveing, G. D., and Dewar, J. *Collected Papers on Spectroscopy.* Cambridge, 1915.

Lockyer, J. N. *Studies in Spectrum Analysis.* New York, 1878.

———. *Chemistry of The Sun.* London, 1887.

Maxwell, J. C. *Theory of Heat.* London, 1870.

Melvill, Thomas. *Physical and Literary Essays.* Edinburgh, 1752.

Niven, W. D., ed. *The Scientific Papers of James Clerk Maxwell.* 2 vols. London, 1890.

Schellen, H. *Spectrum Analysis.* Translated by Jane and Caroline Lassell. New York, 1872.

Smyth, R. Angus, ed. *Chemical and Physical Researches by Thomas Graham.* Edinburgh, 1876.

Tait, P. G. *Lectures on Some Recent Advances in Physical Science with a Special Lecture on Force.* 3rd ed. London, 1885.

[Tait, P. G., and Stewart, Balfour.] *The Unseen Universe.* New York, 1875.

Thomson, J. J. *A Treatise on the Motion of Vortex Rings.* London, 1883.

Thomson, Sir J. J. *Recollections and Reflections.* New York, 1937.

Watson, H. W. *A Treatise on the Kinetic Theory of Gases.* 2nd ed. Oxford, 1893.

Wurtz, A. *La Théorie Atomique.* Paris, 1872.

PAPERS

Alter, David. On certain physical properties of the light of the electric spark, within certain gases as seen through a prism. *American J. Sc.* 19 (1855): 213–14.

Ames, J. S. Grünwald's mathematical spectrum analysis. *Nature* 40 (1889): 19.

———. On the relations between the lines of various spectra, with special reference to those of cadmium and zinc, and a redetermination of their wavelengths. *Phil. Mag.* 30 (1890): 33–48.

Ångstrom, A. Optiska undersökningar. [1853] *Konliga Svenska Vetenskaps Akademiens Handligar* (1852): 229–32; in *Phil. Mag.* 9 (1855): 327–42.

———. Sur les spectres des gaz simples. *Comptes Rendus* 73 (1871): 369–73.

Ångstrom, A., and Thalén, R. Recherches sur les spectres des métalloids. *Nova Acta Regiae Societatis Scientarium Upsalensis* 9 (1875), Article 9 (34 pp.), p. 5.

Balmer, J. J. Notiz ueber die Spectrallinien des Wasserstoffs. *Verhandlungen der Naturforschenden Gesellschaft in Basel* 7 (1885): 548–60, 750–52; and *Ann. der Physik* 25 (1885): 80–87.

———. A new formula for the wave-lengths of spectral lines. *Astrophysical J.* 5 (1897): 199–209.

Becquerel, E. Report on the researches of M. Arn. Thenard concerning the actions of electric discharges upon gases and vapours. *Phil. Mag.* 45 (1873): 154–56.

Berthelot, M. [Sur l'existence réelle d'une matière monoatomique.] *Comptes Rendus* 82 (1875): 1129–30.

Boltzmann, L. Ueber die mechanische Bedeutung des zweiten Hauptsatzes der Wärmetheorie. *Berichte Akad. Wissen. Wien* 53 (1866): 195–220.

———. Ueber das Wärmegleichgewicht zwischen mehratomigen Gasmolekülen. *Berichte Akad. Wissen. Wien* 63 (1871): 397–418.

———. Ueber die Natur der Gasemoleküle. *Berichte Akad. Wissen. Wien* 74 (1876): 553–60.

———. On certain questions of the theory of gases. *Nature* 51 (1894–95): 413–15.

Bosanquet, R. The velocity of sound, and ratio of specific heats, in air. *Phil. Mag.* 3 (1877): 271–78.

Brewster, D. Description of a monochromatic lamp for microscopical purposes, etc., with remarks on the absorption of the prismatic rays by coloured media. *Trans. Roy. Soc. Edinburgh* 9 (1823): 433–44.

———. Observations on the absorption of specific rays, in reference to the undulatory theory of light. *Phil. Mag.* 2 (1832): 360–63.

———. Observations on the lines of the solar spectrum, and on those produced by the earth's atmosphere, and by the action of nitrous acid gas. *Trans. Roy. Soc. Edinburgh* 12 (1834): 519–30.

———. On the luminous bands in the spectra of various flames. *Report British Assn.* 11 (1842): part 2, 15–16.

———. On the action of two blue oils upon light. *Report British Assn.* 12 (1843): part 2, 8.

———. Report on the recent progress in optics. *Report British Assn.* 1 (1831–32): 308–22.

Brewster, D., and Gladstone, J. H. On the lines of the solar spectrum. *Phil. Trans.* 150 (1860): 149–60.

Brillouin, M. Sur le degré de complexité des molecules gazeuses. *Comptes Rendus* 112 (1891): 575–77.

Brodie, Sir B. C. An experimental inquiry on the action of electricity on oxygen. *Phil. Trans.* 162 (1872): 435–84.

Bryan, G. H. Professor Boltzmann and the kinetic theory of gases. *Nature* 51 (1894–95): 21.

Bunsen, R. Ueber Benutzung der Flammenspektren bei der chemischen Analyse. *Verhandlungen des Naturhistorisch-Medicinischen Vereins zu Heidelberg* (1859–60): 31–32.

Bunsen, R., and Kirchhoff, G. Chemische Analyse durch Spectralbeobachten. *Ann. der Physik* 110 (1860): 160–89; 113 (1861): 337–425. (Translated in *Phil. Mag.* 20 [1860]: 89–109; 22 [1861]: 329–49, 498–510.)

Bunsen, R., and Roscoe, H. E. Photo-chemical researches. Part 1. Measurement of the chemical action of light. *Phil. Trans.* 147 (1858): 355–80.

Burbury, S. H. The Ratio of the Specific Heats of Gases. *Nature* 51 (1894–95): 127.

Cartmell, R. On a photochemical method of recognizing the non-volatile alkalies and alkaline earths. *Phil. Mag.* 16 (1858): 328–33.

Ciamician, G. L. Ueber die Spectren der chemischen Elemente und ihrer Verbindungen. *Sitzungsberichte Akad. Wissen. Wien* 76 (1877): Abt. 2, 499–517.

———. Spectroskopische Untersuchungen. *Sitzungsberichte Akad. Wissen. Wien* 82 (1880): Abt. 2, 425–27.

Clausius, R. The nature of the motion we call heat. *Phil. Mag.* 14 (1857): 108–27.

———. On the mean lengths of the paths described by the separate molecules of gaseous bodies. *Phil. Mag.* 17 (1859): 81–91.

Clifton, R. B. An attempt to refer some phenomena attending the emission of light to mechanical principles. *Proc. Literary and Philosophical Society, Manchester* 5 (1866): 24–28.

Cornu, A. Sur les raies spectrales spontanément renversables et l'analogie de leurs lois de répartition et d'intensité avec celles des raies de l'hydrogène. *Comptes Rendus* 100 (1885): 1181–85.

———. On the distinction between spectral lines of solar and terrestrial origin. *Phil. Mag.* 22 (1886): 458–63.

Culverwell, E. P. The kinetic theory of gases. *Nature* 51 (1894–95): 78–79.

Deslandres, H. Spectre du pôle négatif de l'azote. Loi générale de répartition des raies dans les spectres des bandes. *Comptes Rendus* 103 (1886): 375–79.

———. Loi de répartition des raies et des bandes, commune à plusieurs spectres de bandes. *Comptes Rendus* 104 (1887): 972–76.

———. Proprieté fondamentale commune aux deux classes des spectres. Characteres distinctifs de chacune des classes. Variations periodique à trois paramètres. *Comptes Rendus* 110 (1890): 748–50.

Despretz, C. Sixième communication sur la pile. Note sur le phénomène chimique et sur la lumière de la pile à deux liquides. *Comptes Rendus* 31 (1850): 418–22.

Dewar, J. (See under G. D. Liveing and J. Dewar)

Diacon, E. De l'emploi du chalumeau à chlor-hydrogène pour l'étude des spectres. *Comptes Rendus* 56 (1863): 653–55.

———. Recherches sur l'influence des éléments électronégatifs sur le spectre des métaux. *Annales de Chime et de Physique* 6 (1865): 5–25.

Dibbits, H. C. Ueber die spectren der flammen einiger gase. *Ann. der Physik* 122 (1864): 497–545.

Ditte, A. Sur les spectres du soufre, du sélénium et du tellure. *Comptes Rendus* 73 (1871): 622–24.

———. Sur les spectres des corps appartenant aux familles de l'azote et du chlore. *Comptes Rendus* 73 (1871): 738–42.

Draper, W. On the production of light by chemical action. *Phil. Mag.* 32 (1848): 100–14.

Dubrunfaut, A. P. Essai d'analyse spectrale appliquée à l'examen des gaz simples et de leurs mélanges. *Comptes Rendus* 69 (1869): 1245–49.

———. Sur les spectres de divers ordres des corps simples. *Comptes Rendus* 70 (1870): 448–51.

Ebert, H. La mécanisme de la luminosité au point de vue de la théorie electromagnétique de la lumière. *Archives des Sciences Physiques et Naturelles* 25 (1891): 489–503.

———. Electrische Schwingungen molecular Gebilde. *Ann. der Physik* 49 (1893): 651–72.

Eder, J. M., and Valenta, E. Ueber die verschiedenen Spectren des Quecksilbers. *Denkschriften der Kaiserlichen Akad. der Wissen. Wien* 61 (1894): 401–30.

Erman, A. Sur la loi de l'absorption de la lumière par les vapeurs de l'iode et du brome. *Comptes Rendus* 19 (1844): 830–37.

Fitzgerald, G. F. Note on Professor Ebert's estimate of the radiating power of an atom, with remarks on vibrating systems giving special series of overtones like those given out by molecules. *Report British Assn.* 61 (1893): 689–90.

———. The kinetic theory of gases. *Nature* 51 (1894–95): 221–22.

———. On some considerations showing that Maxwell's theorem of equal partition of energy among the degrees of freedom of atoms is not inconsistent with the various internal movements exhibited by the spectra of gases. *Nature* 51 (1894–95): 452–53.

Forbes, J. Note relative to the supposed origin of the deficient rays in the solar spectrum; being an account of an experiment made at Edinburgh during the annular eclipse of 15th May, 1836. *Phil Trans.* 126 (1836): 453–55.

Fraunhofer, J. von Bestimmung des Brechungs- und Farbenzerstreuungs-Vermögens verschiedener Glasarten, in Bezug auf die Vervollkommung achromatischer Fernröhre. *Denkschriften der Königlichen Akademie der Wissenschaften zu München* 5 (1814–15): 193–226.

———. Kurzer Bericht von den Resultaten neurer Versuche ueber die Gesetze des Lichtes, und die Theorie derselben. *Ann. der Physik* 74 (1823): 337–78.

Galitzin, F. B. Zur Theorie der Verbreiterung der Spectrallinien. *Ann. der Physik* 56 (1895): 78–99.

Goldstein, E. Ueber Beobachtungen an Gasspektris. *Ann. der Physik* 154 (1875): 128–49.

———. Ueber das Bandenspectren der Luft. *Ann. der Physik* 15 (1882): 280–88.

Graham, T. Speculative ideas respecting the constitution of matter. *Phil. Mag.* 27 (1864): 81–84.

Grünwald, A. K. Ueber die merkwurdigen Beziehungen zwischen dem Spektrum des Wasserdampfes und den Linienspectren des Wasserstoffs und Sauerstoffs, sowie über die chemische Structur der beiden letzten und ihre Dissociation in der Sonnenatmosphäre. *Astronomische Nachrichten* 117 (1887): 201–14.

———. Mathematische Spectralanalyse des Magnesiums und der Kohle. *Berichte Akad. Wissen. Wien* 96 (1888): 1154–1216.

Handl, A. Notiz ueber die absolute Intensität und Absorption des Lichts. *Berichte Akad. Wissen. Wien* 64 (1872), 2: 129–32.

Hartley, W. N. On homologous spectra. *J. Chem. Soc. London* 43 (1883): 390–400.

Hartley, W. N., and Huntington, A. K. Researches on the action of organic substances on the ultra-violet rays of the spectrum. *Phil. Trans.* 170 (1879): part 1, 257–74.

Helmholtz, H. von. Ueber Integrale der hydrodynamischen Gleichungen welche den Wirbelbewegungen entsprechen. *Crelle's J.* 55 (1858): 25–55.

Herschel, A. S. On a relation between the spectrum of hydrogen and acoustics. *Observatory* 19 (1896): 232–34.

Herschel, J. W. F. On the absorption of light by coloured media, and on the colours of the prismatic spectrum exhibited by certain flames; with an account of a ready mode of determining the absolute dispersive power of any medium, by direct experiment. *Trans. Roy. Soc. Edinburgh* 9 (1823): 445–60.

————. On the absorption of light by coloured media, viewed in connection with the undulatory theory. *Phil. Mag.* 3 (1833): 401–12.

Herschel, W. On the power of penetrating into space by telescopes. *Phil. Trans.* 90 (1800): 49–85.

Hicks, W. On the steady motion of a hollow vortex. *Proc. Roy. Soc.* 35 (1883): 304–8.

————. On the problem of two pulsating spheres in a fluid. *Proc. Cambridge Phil. Soc.* 3 (1880): 276–85.

Huggins, W. On the photographic spectra of stars. *Phil. Trans.* 171 (1880): 669–90.

Julius, V. A. Sur les spectres des lignes des éléments. *Annales de L'École Polytechnique de Delft* 5 (1889): 1–117.

Kayser, H. Ueber Grünwald's mathematische Spectralanalyse. *Chemiker Zeitung* 13 (1889): 1655; 1687–89.

————. On the spectrum of ζ Puppis. *Astrophysical J.* 5 (1897): 95–96.

————. On the spectrum of hydrogen. *Astrophysical J.* 5 (1897): 243.

————. Ueber die Bogenspectren der Elemente des Platingruppe. *Abhandl. Berlin Akad.* (1897).

Kayser, H., and Runge, C. *Ueber die Spectren der Elemente.* 7 parts. Berlin, 1889–93.

————. Ueber die Spectren der Alkalien. *Ann. der Physik* 41 (1890): 302–20.

Kirchhoff, G. Ueber das Sonnenspectrum. *Verhandlungen des Naturhistorisch-Medicinschen Vereins zu Heidelberg* (1857–59): 251–55.

————. Ueber die Fraunhofer'schen Linien. *Monatsberichte Akad. Wissen. Berlin* (1859): 662–65.

————. Contributions towards the history of spectrum analysis and of the analysis of the solar atmosphere. *Phil. Mag.* 25 (1863): 250–62. (Also, see under R. Bunsen and G. Kirchhoff.)

Koláček, F. Versuch einer Dispersionserklarung vom Standpunkte der electromagnetischen Lichttheorie. *Ann. der Physik* 32 (1887): 224–55.

————. Ueber electrische Oscillationen in einer leitenden und polarisationsfähigen Kugel. Ein Beitrag zur Theorie der Spectra einfachster Beschaffenheit. *Ann. der Physik* 58 (1896): 271–310.

Kundt, A., and Warburg, E. Ueber die specifische Wärme des Quecksilbersgases, *Ann. der Physik* 157 (1876): 353–69.

Larmor, J. A dynamical theory of the electric and luminiferous medium. *Phil. Trans.* 185 (1894): 719–822.

———. A dynamical theory of the electric and luminiferous medium. Part II: Theory of electrons. *Phil. Trans.* 186 (1895): 695–743.

Lecoq de Boisbaudran, F. Sur la constitution des spectres lumineux. *Comptes Rendus* 69 (1869): 445–51; 606–15; 694–700.

Liveing, G. D., and Dewar, J. On the spectra of sodium and potassium. *Proc. Roy. Soc.* 29 (1879): 398 ff.

———. On the spectra of magnesium and lithium. *Proc. Roy. Soc.* 30 (1880): 93–99.

———. Investigations on the spectrum of magnesium. No. 1. *Proc. Roy. Soc.* 32 (1881): 189–203.

———. On the identity of spectral lines of different elements. *Proc. Roy. Soc.* 32 (1881): 225–30.

———. On the ultra-violet spectra of the elements. Part I: Iron (with a map). Part II: Various elements other than iron (K, Na, Li, Ba, Sr, Ca, Zn, Au, Tl, Al, Pb, Sn, Sb, Bi, Si, C). *Phil. Trans.* 174 (1883): 187 ff.

———. Notes on the absorption-spectra of oxygen and some of its compounds. *Proc. Roy. Soc.* 46 (1889): 222–30.

Lockyer, J. N. Spectroscopic notes. No. 2: On the evidence of variation in molecular structure. *Proc. Roy. Soc.* 22 (1873–74): 372–74.

———. Researches in spectrum-analysis in connection with the spectrum of the sun. No. 3. *Phil. Trans.* 164 (1874): 479–94.

———. Atoms and molecules spectroscopically considered. *Nature* 10 (1874): 69–71, 89–90.

———. Preliminary note on the compound nature of the line spectra of elementary bodies. *Proc. Roy. Soc.* 24 (1876): 352.

———. Researches in spectrum analysis in connection with the spectrum of the sun. No. 7. Discussion of the working hypothesis that the so-called elements are compound bodies. *Proc. Roy. Soc.* 28 (1879): 157–80.

———. On the necessity for a new departure in spectrum analysis. *Nature* 21 (1879–80): 5–8.

———. On a new method of spectrum observation. *Proc. Roy. Soc.* 30 (1880): 22–31.

———. On multiple spectra. *Nature* 22 (1880): 4–7, 309–12, 562–65.

———. Solar physics: Chemistry of the sun. *Nature* 24 (1881): 267–74, 296–301, 315–24, 365–70, 391–99.

———. On the chemistry of the hottest stars. *Proc. Roy. Soc.* 61 (1897): 148–209.

Lodge, O. On the connection between aether and matter. *Report British Assn.* 61 (1893): 688.

Lommel, E. Ueber Fluoreszenz. *Ann. der Physik* 143 (1871): 26–51.

Loschmidt, J. Schwingungen einer elastichen Hohlkugel. *Sitzungsberichte Akad. Wissen. Wien* 93 (1886): 434–36.

Mascart, E. Recherches sur le spectre solaire ultra-violet et sur la determination des longuers d'onde. *Annales Scientifiques de L'École Normale Supérieure* 1 (1864): 219–62.

———. Sur les spectres ultra-violets. *Comptes Rendus* 69 (1869): 337–38.

———. Recherches sur la determination des longuers d'onde. *Annales Scientifiques de L'École Normale Supérieure* 4 (1867): 1–31.

Masson, A. Études de photométrie électrique. *Annales de Chimie et de physique* 14 (1845): 129–95; 30 (1850): 5–55; 31 (1851): 295–326; 45 (1855): 385–454.

Maxwell, J. C. Illustrations of the dynamical theory of gases. *Phil. Mag.* 19 (1860): 19–32; 20 (1860): 21–37.

———. On the dynamical theory of gases. *Phil. Trans.* 157 (1867): 49–88.

———. Molecules. *Nature* 8 (1873): 437–41.

———. On the dynamical evidence of the molecular constitution of bodies. *Nature* 11 (1875): 357–59, 374–77.

———. "Atom" in *Encyclopaedia Britannica. 9th ed.* Edinburgh, 1875–89, 2 (1875).

———. The kinetic theory of gases. *Nature* 16 (1877): 242–46.

Miller, W. A. Experiments and observations on some cases of lines in the prismatic spectrum produced by the passage of light through coloured vapours and gases, and from certain coloured flames. *Phil. Mag.* 27 (1845): 81–91.

———. On the photographic transparency of various bodies, and on the photographic effects of metallic and other spectra obtained by means of the electric spark. *Phil. Trans.* 152 (1862): 861–87.

———. On spectrum analysis. *Chemical News* 5 (1862): 201–3, 214–18.

Mitscherlich, A. Beitrage zur Spectralanalyse. *Ann. der Physik* 116 (1862): 499–507.

———. Ueber die spectren der verbindungen und der einfachen korper. *Ann. der Physik* 121 (1864): 459–88.

Moser, J. Die Spectren der chemischen Verbindungen. *Ann. der Physik* 160 (1877): 177–99.

Mosotti, O. F. On the forces which regulate the internal constitution of bodies. *Scientific Memoirs,* edited by Richard Taylor, 1 (1837): 448–69.

Nordenskiöld, A. E. Sur un rapport simple entre les longuers d'onde des spectres. *Comptes Rendus* 105 (1887): 988–95.

Pickering, E. C. Stars having peculiar spectra. New variable stars in Crux and Cygnus. *Astrophysical J.* 4 (1896): 369–70.

———. The spectrum of ζ Puppis. *Astrophysical J.* 5 (1897): 92–94.

Plücker, J. Ueber die Einwirkung des Magneten auf die electrische Entladung in verdünnten Gasen. *Ann. der Physik* 104 (1858): 113–28; 105 (1859): 67–84.

———. Analyse spectrale. *Cosmos* 21 (1862): 283–88; 312–15.

Plücker, J., and Hittorf, W. On the spectra of ignited gases and vapours, with especial regard to the different spectra of the same elementary gaseous substance. *Phil. Trans.* 155 (1865): 1–29.

Powell, Baden. On the analogies of light and heat. *Proc. Royal Inst.* 1 (1851–54): 172–78.

Preston, S. T. On the physical aspects of the vortex-atom theory. *Nature* 22 (1880): 56–59.

Rayleigh, Lord. On the propagation of waves along connected systems of similar bodies. *Phil. Mag.* 44 (1897): 356–62.

Roberts-Austen, W. C. Metals at high temperatures. *Proc. Roy. Inst.* 13 (1892): 502–18.

Roscoe, H. E. On Bunsen and Kirchhoff's spectrum observations. *Notice of the Proceedings of the Royal Institution* 3 (1858–62): 323–28.

Roscoe, H. E., Clifton, R. B. On the effect of increased temperature upon the nature of the light emitted by the vapour of certain metals or metallic compounds. *Proc. Lit. Phil. Soc. Manchester* 2 (1860–62): 227–30; and *Chemical News* 5 (1862): 233–34.

Runge, C. On the harmonic series of lines in the spectra of the elements. *Report British Assn.* 56 (1888): 576–77.

———. On the line spectra of the elements. *Nature* 45 (1891–92): 607–8; 46 (1892): 100, 200, and 247.

(Also, see under H. Kayser and C. Runge)

Rydberg, J. R. Om de kemiska grundämnenas periodiska system. *Bihang till Konliga Svenska Vetenskaps-Akademiens Handligar* 10 (1884–85), no. 2, 31 pp.

———. On the structure of the line spectra of the chemical elements. *Phil. Mag.* 29 (1890): 331–37; and *Comptes Rendus* 110 (1890): 394–97.

———. Recherches sur la constitution des spectres d'émission des éléments chimiques. *Konliga Svenska Vetenskaps-Akademiens Handligar* 23 (1888–89), no. 11, 155 pp.

———. The new elements of clevite gas. *Astrophysical J.* 4 (1896): 91–96.

———. The new series in the spectrum of hydrogen. *Astrophysical J.* 6 (1897): 233–38.

———. On triplets with constant differences in the line spectrum of copper. *Astrophysical J.* 6 (1897): 239–43.

———. On the constitution of the red spectrum of argon. *Astrophysical J.* 6 (1897): 338–48.

———. La distribution des raies spectrales. *Rapports présentés au Congrès International de Physique reuni à Paris en 1900.* Paris, 1900, 2: 220–24.

Salet, G. Sur le spectre d'absorption de la vapeur de soufre. *Comptes Rendus* 73 (1871): 559–61; and 74 (1872): 865–66.

———. Sur les spectres du sélénium et du tellure. *Comptes Rendus* 73 (1871): 742–45.

———. Sur le spectre primaire de l'iode. *Comptes Rendus* 75 (1872): 76–77.

———. Sur les spectres des métalloids. *Annales de Chimie et de Physique* 28 (1873): 5–71.

———. Sur les spectres multiples. *Association Française pour l'Avancement des Sciences, Comptes Rendus* (1875): 225–27.

———. Sur les spectres doubles. *J. de Physique* 4 (1875): 225–27.

————. Sur le spectre d'azote et sur celui des métaux alcalins, dans les tubes de Geissler. *Comptes Rendus* 82 (1876): 223–26; and *J. de Physique* 5 (1876): 95–97.

Schuster, A. On the spectrum of nitrogen. *Proc. Roy. Soc.* 20 (1872): 484–87; and *Phil. Mag.* 44 (1872): 537–41.

————. On the spectrum of hydrogen. *Nature* 6 (1872): 358–60.

————. Spectrum of nitrogen. *Nature* 8 (1873): 161.

————. Researches on the spectra of metalloids. *Nature* 15 (1877): 401–2.

————. Spectra of metalloids. *Nature* 15 (1877): 447–48.

————. The spectra of chemical compounds. *Nature* 16 (1877): 193–94.

————. On the spectra of metalloids. Spectrum of oxygen. *Proc. Roy. Soc.* 27 (1878): 383–88; and *Phil. Trans.* 170 (1879): 37–54.

————. On harmonic ratios in the spectra of gases. *Nature* 20 (1879): 533.

————. On our knowledge of spectrum analysis. *Report British Assn.* 48 (1880): 258–98.

————. On harmonic ratios in the spectra of gases. *Proc. Roy. Soc.* 31 (1881): 337–47.

————. The teachings of modern spectroscopy. *Popular Scientific Monthly* 19 (1881): 466–82.

————. The genesis of spectra. *Report British Assn.* 50 (1882): 120–43.

————. The kinetic theory of gases. *Nature* 51 (1894–95): 293.

————. On a new law connecting the periods of molecular vibrations. *Nature* 55 (1896–97): 200–1.

————. Professor C. Runge and F. Paschen's researches on the spectra of oxygen, sulphur and selenium. *Nature* 57 (1897–98): 320–21.

————. On the chemical constitutions of stars. *Proc. Roy. Soc.* 61 (1897): 198–213.

Secchi, P. A. Sur la constitution de l'auréole solaire, et sur quelques particularitiés offertes par les gaz raréfiés, lorsqu'ils sont rendus incandescent par les courants électriques. *Comptes Rendus* 70 (1870): 79–84.

Simon, C. Sur le rapport des deux chaleurs specifiques. *Comptes Rendus* 83 (1876): 726–28.

Smyth, C. P. Micrometrical measures of gaseous spectra under high dispersion. *Trans. Roy. Soc. Edinburgh* 32 (1887): 415–60.

Soret, J. L. On harmonic ratios in spectra. *Phil. Mag.* 42 (1871): 464–65.

Stearn, C. H. Spectrum of nitrogen. *Nature* 7 (1873): 463.

Stewart, B. An account of some experiments on radiant heat, involving an extension of Prevost's law of exchanges. *Report British Assn.* 27 (1858): part 2, 23–24.

Stokes, G. G. On the change of refrangibility of light. *Phil. Trans.* 142 (1852): 463–562.

————. On the long spectrum of electric light. *Phil. Trans.* 152 (1862): 599–619.

Stoney, G. J. The internal motions of gases compared with the motions of waves of light. *Phil. Mag.* 36 (1868): 132–41.

————. On the cause of the interrupted spectra of gases. *Phil. Mag.* 41 (1871): 291–96.

———. On the cause of double lines and of equidistant satellites in the spectra of gases. *Trans. Royal Dublin Soc. Sc.* 4 (1888–92): 563–608.

Stoney, G. J., and Reynolds, J. E. An inquiry into the cause of the interrupted spectra of gases. Part II. On the absorption-spectrum of chlorochromic anhydride. *Phil. Mag.* 42 (1871): 41–52.

Swan, W. On the prismatic spectra of the flames of compounds of carbon and hydrogen. *Trans. Roy. Soc. Edinburgh* 21 (1857): 411–30.

———. Note on Professors Kirchhoff and Bunsen's paper "On chemical analysis by spectrum observations." *Phil. Mag.* 20 (1860): 173–75.

Talbot, W. H. F. Some experiments on coloured flames. *Edinburgh J. Sc.* 5 (1826): 77–81.

———. Facts relating to optical science. No. 2. *Phil. Mag.* 4 (1834): 112–14.

———. On the nature of light. *Phil. Mag.* 7 (1835): 113–18.

———. Facts relating to optical science. No. 3. *Phil. Mag.* 9 (1836): 1–4.

———. Note on the early history of spectrum analysis. *Proc. Roy. Soc. Edinburgh* 7 (1869–72): 461–66.

Thomson, J. J. On the cathode rays. [1897] *Proc. Cambridge Phil. Soc.* 9 (1898): 243–44.

———. Cathode rays. *Proc. Roy. Inst.* 15 (1897): 419–32.

Thomson, W. On vortex atoms. *Proc. Roy. Soc. Edinburgh* 6 (1867): 94–105.

———. Vibrations of a columnar vortex. *Phil. Mag.* 10 (1880): 155–68.

———. Vortex statics. *Proc. Roy. Soc. Edinburgh* 9 (1878): 59–73.

Troost, L., and Hautefeuille, P. Sur les spectres du carbone, du bore, du silicium, du titane et du zirconium. *Comptes Rendus* 73 (1871): 620–22.

Villarceau, A. Yvon. Note sur les determinations théorique et expérimentale du rapport des deux chaleurs spécifiques dans les gaz parfaits dont les molecules seraient monoatomiques. *Comptes Rendus* 82 (1876): 1127–28.

Vogel, H. W. Ueber die photographischen Aufnahme von Spectren in Geissler-röhren eingeschlossenen Gase. *Monatsberichte Akad. Wissen. Berlin* (1879): 115–19.

———. Ueber die Spectra des Wasserstoffs, Quecksilbers und Stickstoffs. *Monatsberichte Akad. Wissen. Berlin* (1879): 586–604.

Watts, W. M. On double spectra. *Quarterly J. Sc.* 1 (1871): 1–15.

Wheatstone, C. On the prismatic decomposition of electric light. (Abstract) *Phil. Mag.* 7 (1835): 299; and *Report British Assn.* 3 (1834), part 2: 11–13. (Original) *Chemical News* 3 (1861): 198–201.

Wiedemann, E. Untersuchungen über die Natur der Spectra. *Ann. der Physik* 5 (1878): 500–24.

———. Ueber das thermische und optische Verhalten von Gasen unter dem Einflusse electrischer Entladung. *Ann. der Physik* 10 (1880): 202–57.

———. Zur Mechanik des Leuchtens. *Ann. der Physik* 37 (1889): 177–248.

Willigen, V. S. M. van der. Ueber das electrische Spectrum. *Ann. der Physik* 106 (1859): 610–32.

Wolf, E., and Diacon, E. Note sur les spectres des métaux alcalins. *Comptes Rendus* 55 (1862): 334–36.

Wollaston, W. H. A method of examining refractive and dispersive powers by prismatic reflection. *Phil. Trans.* 92 (1802): 365–80.

Wrede, F. J. von. Attempt to explain the absorption of light according to the undulatory theory. *Scientific Memoirs,* edited by Richard Taylor 1 (1837): 477–82, 483–502.

Wüllner, A. Ueber die Spectra einiger Gase in Geissler'schen Röhren. *Ann. der Physik* 135 (1868): 497–527.

———. Sur les spectres des gaz simples. *Comptes Rendus* 70 (1870): 125–29.

———. Ueber die Spectra einiger Gase in Geissler'schen Röhren. *Ann. der Physik* 144 (1872): 481–525.

———. Ueber die Spectra der Gase in Geissler'schen Röhren; die Entstehung der Spectra verscheidener Ordnung. *Ann. der Physik* 147 (1872): 321–53.

———. Ueber die Spectra der Gase in Geissler'schen Röhren. *Ann. der Physik* 149 (1872): 103–12, 321–53.

———. Einige Bemerkungen zu Herr Goldstein's Beobachtungen an Gasspectris. *Monatsberichte Berlin Akad.* (1874): 755–61.

Young, C. A. Spectroscopic Notes, 1879–1880. *American J. Sc.* 20 (1880): 353–58.

Young, T. On the theory of light and colours. *Phil. Trans.* 92 (1802): 12–48.

Zöllner, F. Ueber den Einfluss der Dichtigkeit und Temperatur auf die Spectra glühender Gase. *Ann. der Physik* 142 (1871): 88–111.

Zeeman, P. On the influence of magnetism on the nature of the light emitted by a substance. *Phil. Mag.* 43 (1897): 226–39.

———. Doublets and triplets in the spectrum produced by external magnetic forces. *Phil. Mag.* 44 (1897): 55–60, 255–59.

Index

Absorption: explanation of, 14, 21, 23, 32, 33; Kirchhoff's principle of, 31; of light by colored media, 29; problem of explanation of, 18; theory of, 20
Acoustical analogy, 22, 105, 111, 112, 121, 126, 127, 131, 133, 166, 177, 199, 200, 202, 204, 205, 206
Adams prize, 173
Air: ratio of specific heats, 162, 163
Alkali metal spectra, 140, 141, 153, 156
Allotropy, 63, 66, 67, 68, 73, 77, 79, 80, 108
Ames, J. S., 101
Ampère, A. M., 44
Ångstrom, A. J.: Fraunhofer lines, 20–22; gaseous and metallic spectra, 13–14; multiple spectra, 59, 62, 64, 65, 66, 68, 71, 72; mentioned, 23, 86, 96, 97, 118, 122, 123, 132, 133, 146
Ångstrom unit, 139
Anti-atomic movement, 204
Astrophysical Journal, 133
Astrophysics, 133, 137, 155
Atomic theory, 8
Atomic weight, 103, 106, 109, 110, 140, 153, 154
Atoms, 78, 79
—Bohr, 194
—Boscovichean, 45, 160, 163, 170, 179
—charged, 82, 185, 186
—chemical: nature of, 78, 185; mentioned, 46, 79, 80, 81, 82, 83, 93, 161, 171, 183, 184, 185, 189, 192
—chemical views of, 35, 45, 46, 47, 78; adopted by physicists, 80

—conditions to be satisfied by, 160
—constitution of, 154, 155
—defined, 176
—ether-coated, 180, 180n, 181, 183, 186, 201
—Lockyer's "true," 79, 85
—Lucretian, 45, 79, 160, 163, 166, 167, 169, 170
—motions of, 154, 155
—nature of, 42, 44, 45, 159, 164, 168, 183
—physical views of, 35, 46, 47, 78, 157; same as chemical, 80
—structure of, 138, 172
—ultimate, 45, 46, 47, 63, 77, 78, 80, 81, 83, 95, 98, 100, 160, 161, 192, 193
—vibrations of, 81
—vortex, 93, 160–61, 163, 165–75, 178, 179, 180, 188, 192, 193, 195, 200, 202, 203; theory of, 93, 162, 165–75 *passim,* 181, 189, 192, 194, 196, 200, 205, 211, 212, 213; theory of, rejected, 182
Avogadro, A., 63

Balmer, J. J., 131–33, 137, 138, 139, 140, 142, 146, 148, 149, 150, 152, 155, 172, 173, 191, 200, 201, 206
Balmer's law. *See* Hydrogen series formula
Bands, dark, in flame spectra, 19
Basic lines: defined, 88; effect of temperature on, 90; explanation of, 89; spurious, 98; mentioned, 87, 91, 97, 100
Basic molecules, 88, 89
Becquerel, E., 116
Bell hypothesis, 96

 THE JOHNS HOPKINS PRESS

Designed by Arlene J. Sheer

Composed in Baskerville text and display
by Service Typographers, Inc.

Printed on 60-lb. Perkins and Squier R
by Universal Lithographers, Inc.

Bound in Holliston Roxite
by L. H. Jenkins, Inc.